EARTH HOLLOW

EXPLORING A FLEXILE, MYSTERIOUS & VERY ALIVE PHYSICAL WORLD

Nicholas C. Eliopoulos

iUniverse, Inc.
Bloomington

EARTH HOLLOW
Exploring A Flexile, Mysterious & Very Alive Physical World

iUniverse books may be ordered through booksellers or by contacting:

iUniverse
1663 Liberty Drive
Bloomington, IN 47403
www.iuniverse.com
1-800-Authors (1-800-288-4677)

ISBN: 978-1-4502-6518-8 (sc)
ISBN: 978-1-4502-6519-5 (e)

Printed in the United States of America

iUniverse rev. date: 9/19/2011

To my great-grandmother on my mother's side,
name unknown to me,
who was illiterate and whose life was confined to the mountain
village, of Kapsia, Arcadia, Hellas.

She is listening to students of the gymnasium returning to the village, of about four hundred people at that time, for holiday. They are trying to explain the real facts behind the eclipses, which had captured the imagination of a local crowd at the churchyard.

Piously crossing herself in disbelief and turning with an awed look to others remaining behind, she utters, "Are they going crazy in the schools? Are they saying that the Earth, from right here under, right here where it is?" She is stomping the ground with one foot as though to prove to herself of its secure flatness and stability under her, "Picks itself up and away? And it goes beyond and in back of the Moon?" She stretches her arm out to her side, as she is motioning as away to the other side of the mountain, which isolates their village. Actually she is pantomiming such unbelievable ethereal motion.

The nation of Hellas had not long ago gained its independence after four centuries of slavery with suppression of education, while West Europe had already been, for four hundred years, blossoming forth with the Renaissance.

This work is also dedicated to those students from the village, which had the good instinct to share their knowledge with the village, often stimulating their minds in the coffee shop. The author also dedicates this to the entire village from which both his parents hail. Since those days, this village has multiplied in numbers and produced university graduates, businesspeople, and professionals living in the five continents.

— —— —

EARTH HOLLOW

ECCENTRIC DYNAMICS

EARTH'S STRUCTURE

EARTH'S PLIABILITY

EARTH'S ENVIRONMENT

PART TWO

MAN'S WORLD

EARTH'S INTERIOR

EARTH'S RELATIVITY

EARTH'S CALENDRICS

EARTH'S SYNCHRONICITY

EARTH'S MINI WORLDS

METAPHYSIC WORLD

QUESTION

Are the planets strong floating bubbles or celestial kites, or are they like lead weights suspended by strong, heavy invisible chains from the Heavens? Do planets bounce off each other, or would they hopelessly shatter upon clashing? Can all such be asked of the stars as well? Is there life inside these celestial ornaments, these planets and stars? Is there an inner little sun suspended in each planet? How is global warming affected? Do planets and stars differ?

Will Man ultimately gain the knowledge to navigate carefully to and pass around these celestial bodies? Can Man deter them from their mathematically arranged courses as divinely ordained; or chaotically so, as the Big Bangists ordain?

Are we talking about naturally made spaceships? Ancient Mythologies suggest yes and no! Can we temporarily bring any other planet and the Earth close enough to moor tangentially at their mutually touching atmospheres and board from one to the other in space? If to any it is yes, the planets in question must be flexile and livable, like the ancient ships at sea were flexible to absorb the shock of giant ocean waves without losing their speed, and to throw the waves harmlessly over to the other side of the ship.

— ——— —

INTRODUCTORY

The Earth's sundry changes dramatically affect the environment of all things, animate and inanimate, including the quality of life for Man.

Mathematics and physics questions do point to a greater possibility for a hollow and flexile Earth than for a solid and rigid one. The delving into the subject at hand was instigated from a study on the issue of the frequently changing measures eluding a more perfect calendar, as well other measurements, such as Eratosthenes calculating the diameter of the Earth: did he fall short of accuracy, or did the Earth since then change in size and configuration? Physical and mathematical questions do not satisfy the current concepts of the natural constituency of the Earth. Historic measurements of time and space are often misunderstood and chalked up as "ancient inaccuracies", allegedly due to their "lack of instruments", as is said. Possibly it is the Earth, which is ever changing in its diameter, rate of axial rotation, magnitude, and eccentricity of orbital travel, spherical eccentricity, and other related motions. To justify the difference in ancient measurements, it would be assumed that the Earth is pliable and elastic rather than brittle and solid as still officially imagined. As to dates of geologic and astronomic happenings none are given, as the work is a theoretic overview.

Hence, this is an adventure into a story about a miraculous *Earth Hollow* which also is –logically– quite flexile.

The information used herein is gleaned from standard encyclopedias, dictionaries, and conventional works on Earth and space studies. This work is synthesized, admittedly, in a speculative manner. The product of synthesizing these thoughts and facts may not be found to be in conformance with government publications, academic papers, and serious authorities whose success leans on their toeing the given line for their survival. Otherwise, of necessity, and for various reasons it is understood that officialdom is obligated to keep faithful to current "dogmato-scientific" tenets ordained from behind the scenes, especially per the little-publicized Solvey Conferences, at least for the time being. These dogmato-scientific tenets are intentionally constructed to appear as immutable universal laws of materialism and the oligarchic New World Order of an alleged New Age (with their agenda and the slate for the ruling robots of the Earth kept secret) to remove Man from his ethics and religion, his tradition of civility, his live contact with the Creator, and from his conscience.

A Bibliography stimulating the author's seriousness would include Dick Tracy, Buck Rogers, Jules Verne, Nikola Tesla, Von Daenicken, Immanuel Velikovsky, Isaac Asimov, Dante, several "pulp" authors, comic book illustrators, and others incidentally, who unjustly may have been overlooked for the depths of their imaginations, or simply forgotten as passing fancies. Long after the passing of most to their reward, not from their reputation for frivolous entertaining, but some for their deepest insights are posthumously upgraded to the status of prophets. And, that honor is not by court rulings, not by academia, but is done through a common witnessing by those commoners enthused by the manifestly obvious through the passing of time, even though as casual entertainment at the time, and nevertheless, within the whole truth about the universe, our grandiose home.

In isolated forums, strange thoughts may not always be barred from being aired out for any clues contributive on the matter of Earth studies. In this adventurous frame of mind, this work is launched. Absolute validity is not insisted upon — rather, it is presented as victuals for contemplating, based on certain given truths, apparent facts, and wandering imagination, as is conventionally perceived for presentation. Suggestions from Greek and other Mythologies, many of which somehow fall into agreement, could make some sense scientifically, are of considerable inspiration, as indeed are the holy Old and New Testaments. The common repetitious sensationalism over the unknown is not made herein into a big issue, but whenever cited, it is as referential matter, as otherwise is done in preparing serious theses for graduate degrees, or better yet, as a detective gathers all imaginable possibilities by which to begin investigating a case.

At this point, it might be opportune to define who is a so-called "practical" person. It is one, whose power of thinking cannot reach beyond the range of his five senses. Or, otherwise, veritably and in a proper sense, a practical person is able to discern what he does not know, and what he does not know yet. On this premise, the undersigned may speak out against the frivolous concepts, and those institutions endangering the well being of this world, rather than edifying it. Where reason fails to be seen or grasped, at times some impersonal ridicule may serve. But patience serves best, even to the next or several generations to take hold.

— —— —

In a piece of steel, gold, wood, bone, flesh, vegetable, seed, juice, anything organic or inorganic, from its visible constituency and knowing that such are subdivided into cells, molecules, atoms, neutrons and on, we see that each of such may be reduced or changed to a state of weightlessness. It is not for the miniscule dimensions, as for practicality, but virtually it counters the reality of attempting to sum up increments to give the total weight of the whole. Subatomic matter manifests as packets of energies with centrifugal qualities and lightning speeds, meaning that they become packets of power energy to make conversions of matter and do things unimaginable, including to overcome gravity.

Therefore, if the infinite universe is an integral entity, all its successive subdivisions, such as the galaxies and astral systems, lead to arranged concentrations of whirling and jumping energies, and attracting inestimable potential for purposes beyond imaginable potential and possibilities. So why can't the Earth float in space as electrons do within an atom? This is mindful of the dispute between Democritus and Anaxagoras on the nature of the infinitesimal and of the nature of the atom, and on the effects of splitting the atom. Axiomatically, we begin by holding that the Earth is hollow and flexile, and we limit the argument accordingly. We mortals find ourselves domiciled somewhere between on the ranging scale between the infinitesimal and the infinite.

— ———— —

For convenience of language, the subjunctive and conditional phraseology, when required for accuracy of statement, at times is relaxed in favor of simple statements in the indicative.

Poetic justice would engender some discretion against putting any of this material into futuristic fictional form. There are no humorous cartoons enveloped with human interactions or reactions, nor coy devices to avoid scorn when anything unorthodox is seriously presented.

This work may serve as a precursor to *Julian Calendar Valid*, by this author. Together the two works, *Julian Calendar Valid* and *Earth Hollow & Flexile,* offer an answer to the nature of global warming and cooling, their affect upon all measurements, and the well being of Man.

Because of the infinite repeating of universal laws and mathematic truths, this work on the Earth extends two ways: from the Earth's surface inward, to its very centroid, and from the Earth's surface outward, to the celestial sphere.

— ———— —

PART ONE

PHYSICAL WORLD

PRECURSORY

Conventional obstacles prejudice our seeing and discussing the truth about the Earth's wandering in the vastness of the universe.

1 — EARTHLY AND ASTRAL FICTION

"Many a true word is spoken in jest."

Many a sound hypothesis can be presented quite safely, that is, when as scientific fiction. People of all classes love volumes of fiction, but when faced with a few controversial facts, some feel burdened, if not threatened. Many feel obligated, as a social duty, to hide behind the masses, as being a part of a united non-committal fantasy and perhaps to protect their sanity.

Generally, it takes scientific fiction, at least, to begin breaking down the cobwebs on fixed ideas, before embarking on a scientific investigation for the truth in any particular matter without prejudice. Especially it is so when something is far from being accepted as politically or academically correct.

In this work, with all sincerity it is attempted to give a hypothetic directions, to establish an approach toward finding some truthful scientific answers to some truth on the nature of the cosmos with its planets and stars, before proceeding toward the greater unknown, which is the Earth from beneath its surface to its very centroid. The work herein is a hypothesis, not a scientific credo, a hypothesis open for discussion and reworking.

Speculative knowledge of the inner unknown world presently is restricted to cutaway sketches, done by professional artists, of a perfectly spheric world with a series of different-colored layers, each perfectly spheric, and indifferent to the existence of poles, as if these layers were accumulated as though the Earth is built up by rolling it around everywhichway, like a billiard ball, or snowball, causing it to accrue the thickened layers uniformly throughout, rather than built distinctively around an axis. It still is as if an axis of rotation had never existed until Columbus discovered America. A scientific impossibility, if the Earth came into being through a snowball process, how could an axis of rotation be justified? How can the fixed rigidity of the location of the axis be justified? However, there is a real possibility for matters to resemble as presented herein.

— ——— —

2 — EARTH CANNOT BE SOLID

One of the most impossible things in the realm of reality is to have a solid Earth, rigid in form and most weighty, and legislate it to maintain suspension in space without skyhooks. That would be illogic, contrary to the very scientific and circumstantial evidence of the science of physics. More and more information and phenomena point toward breaking from some established so-called "scientific" attitudes. By now old tenets might well be stale, and timely pointing toward hypothesizing that the Earth is hollow, and that there are succinct laws maintaining the Earth's invisible suspension in the Heavens, as well as that of all the heavenly bodies.

As part of the hypothesis, the hollow of the Earth is not to be imagined as an arbitrary formless or difficult-to-define "murky" void – or as having murky vast tunnels carved out by super-worms and waters rushing through, or being mainly composed of hollows like a sponge. It is molded into a definite structured shape through certain celestial and internal forces. There are many natural phenomena, which are not explainable when seen on the basis of a solid Earth, and in general of solid planets, and yet these are flamboyantly ignored. After all, but in deference to the doubters, yes, ingenious Man's heavy machines, giant thumping dinosaurs, and galloping elephants have been stomping the Earth's surface harder and harder for ages, and so what.

We understand soap bubbles staying in suspension in an enclosed space for a few minutes, or those outdoors fleeing up and away, and we understand why marbles and golf balls are not able to do so. We accept, after the demonstrated fact, that the heavy flying machines do fly. We accept that kites, light as they are, are heavier than air, yet they can sustain themselves in air un-propelled. The argument

3

is in the air currents. Bravo! We are catching on; are there other unknown currents, too?

On the contrary, we accept a belief that solid planetary masses are freely, orderly, and invisibly suspended in the Heavens —that like solid lead balloons, they are expected to float through the vastness of space— but we scoff at the possibility of their being hollow like bubbles. Why? Because nobody id supposed to believe such a thing!

We accept the notion, that the rhythmic timing and celestial trajectories of these heavenly bodies allegedly are fixed and exact, and repeating so, throughout the millennia. And oppositely, we scoff at the possibility that all motions, paths, and timing cycles are never the same, for that, too, hurts our security and reputation for cultural promptness; how can these be different each time? But all these have a degree of elasticity and consequently vary with each successive cycle, so as not to disrupt the rhythm to which they are subject. Astronomers continuously struggle in vain for more accurate measurements. We cannot believe that the sky is different each day and night as we keep seeing it; and that never in the ages will the same sky will reappear to us. Nevertheless, the sky is as endlessly varying as a kaleidoscope.

Man's absolutistic thinking is part of the problem, his being totally either an inflexible "creationist" or a "straight-liner" evolutionist. Careful study of Holy Writ in the Hellenic language offers clues to the etymology of words employed, to break across the errantly assumed semantics of later linguistic developments, by which retroactively to apply seriously to certain Bible studies outside the Hellenic language.

Inconsistently, however, we most readily do accept the Chinese proverb "We never see the same river twice", perhaps because water is wet and flows, right before our very seeing eyes, with a pattern of leaves and debris never the same. Yet most do not quite fathom the depth of the universally applicable classic Greek proverb that "everything is in a state of flux," or "ta panta rhei", simply meaning that "everything flows".

– ––––– –

3 — PSYCHOLOGIC INERTIA

In the primordial Golden Age of mankind, before the beginning of any presently known ancient civilizations, before Man developes the need for keeping records, before the evolution of writing, when Man is at peace, he keeps a well-intact oral tradition metered into poetic form, song, dance, and pantomime, among so many other cultural memories. Among them, the Earth is round, and that it sails around the Heavens, while continuously wheeling around on its own axis. The evidence occurs in the analysis of many Mythologies and ancient folklores, as well in the Old Testament (Septuagint), all of which had been reduced to writing much later in antiquity.

With the advent of historic times, accompanied with Man's losing knowledge and control of his environment, and with the rise of violence and its sister, insecurity, as Man becomes ethically and psychologically lost in his own environment, the Earth begins to appear as a two-dimensional plane – flat. Man imagines he could escape to anywhere on the wide flat Earth, or get lost upon it in circles, in order to survive, but he has to be careful not to approach the dreadful "edge" of the Earth and fall off. And fall to where? He dares not to answer, for secretly he harbors that ineffable fear. Is this one of such syndromes, fear, that keeps him from thinking?

In ancient Egypt, it is said, that a brave admiral circumnavigates Africa. Upon returning to Egypt on her other coast, he is sentenced to death for sacrilege, for what he did is religiously impossible, a taboo. Sparing his life would force hubris upon their religio-social values. Had there been another coexisting religion offering physical security to the populace of Egypt, and keeping her immune against such fears of reality, such an expiable death sentence might not have been necessary. However, history remembers this martyr of reality of truth, Admiral Hanno.

In today's twenty-first century, even among the civilized nations, still there are hosted the creedsmen of a flat-Earth society. By both the scions of the new scientific tradition, as well as the more astute laboring layman, the "flat-earthists" are humorously considered an oddity; and their feelings and reasoning are intriguing, amusing, and perhaps sophisticatedly challenging. One wonders about those bankers, religionists, and favored promising political hacks seeking control of the world. Do they realize where they fit in the universe they seek to control?

Actual experience with advanced methods of observation, not the power of reason alone, lead most of the literate of humanity not to take any flat-earth society too seriously. It is no longer in vogue. The flat-Earth believers by default, however, do insinuate there is a hollowness, an abysmal void, relating to the Earth. If the Earth were a giant carpet, what could be under the carpet? How thick, allegedly, is the terrestrial carpet?

Today, the Earth is round again, a new day for inducing mankind to start coming out of a medieval setting. But to satisfy certain human fears and emotional needs, it imperatively is —"It just *has* to be," they say— a solid and rigidly shaped round Earth of hard durometer and more solid than a shatter-proof billiard ball. Scientific knowledge alone would be insufficient and futile to convince otherwise the honest masses and esteemed intellectuals. For such knowledge would violate or fail to satisfy some current psychologic, sectarian, moral, or social need for our bodily security on our planet. Because Man (the individual with his-her fellows) nourishes a need for experiencing a sense of solidity under his feet. A stabile base under his feet is a required credo for him, in spite of the incoherence in his senses as to the rest of the universe.

In the Great Pyramid, built early in Egypt's historical period, the builders of which remain unknown, there are measurements that give accurate evidence of ancient knowledge of a round Earth, along with a related complex of three-dimensional measurements of celestial bodies far into space and their pertinent data, revealing the Earth's astronomical relationships (Flinders Petrie) with them. The pyramids also establish the measurements and corresponding multifaceted data of the other planets. (See: Hippokrates Dakoglou,

On Pythagoras, in Greek). The locations of many prehistoric Pagan temple sites throughout Europe and North Africa form a triangular grid. The location of the Great Pyramid, however, has slipped from the grid, indicating that Africa has slightly moved northward at the expense of the Mediterranean Sea.

As West Europeans 500 years ago are circumnavigating the Earth, concurrently with the invention of the printing press, the round Earth belief becomes instilled in the manner of romantic fiction, then as straight and overpowering propaganda, and accordingly the shrewd bankers proceed to finance empires to grab all they could through them, on our finite round world. Then all these beliefs become universally acceptable and are finally ordained degreed sciences.

However, by now, Man has "progressed" to being preoccupied with many affairs and thoughts about his accumulating personal wealth in registered, deposited, and sacredly secured printed "numbers" ($) on pieces of paper —products of trees (paper) and octopuses (ink)— and is preoccupied with his materialistic "rights and privileges" and his personal deification (rather than saving his honor, soul, and his culturo-religious security).

Perhaps with the invention of the aeroplane, wherein one's feet are not rested on a solid foundation, and nevertheless one feels safe being carried therein, floating in air above a round Earth and looking out a shatterproof window, he sees the wings of his "flying carpet". Albeit of necessity, the Earth convincingly is still solid. After its temporary suspension in space, the aeroplane returns him to land upon his most sure-footed solid Earth, with a dramatic sigh.

The cinematograph serves well for the masses of viewers in having Man experience that the Earth is a ball invisibly "suspended" in space, as is stated in a psalm of David. One can have his feet firmly planted on the solid concrete floor inside an immobile and sound earthquake-proof building housing the theater, while on the motion picture screen, he is sensually, dramatically, and like a feather carried off into space by an airplane of advanced technology to look down upon the great curvature of his Earth, his home.

Man is no longer threatened psychologically, and so he accepts in his mind, heart, gut, and through a sensual experience, although by proxy, that the Earth must be a sphere, unquestionably not flat, never was flat. A flat Earth no longer offers any needed psychologic security. The movie houses, most of all, dramatize it as proven not flat, impossible to be flat, with no further proof needed. They see it with their very own eyes (in the movies), certainly the world is round.

— ——— —

4 — PSYCHOPATHIC HANGOVER

At the time when a part of ancient and medieval mankind envisaged the world as flat, naturally he envisioned limits edging the surface. But, as for the depth of his Earth, he was not seriously concerned, depth not being a part of his two-dimensional mentality. He never even dreamed of digging a well too deep to pierce his secure sheet of earth. Even so, what a most secure feeling he embosoms in his fantasy of the unknown! A more speculative individual inside a well might take precautions, bracing himself to the sides lest the well lose its bottom, more suddenly than falling off the edge.

Without relenting on the psychological need for a solid foundation, at first it is unimaginable that a newly accepted round Earth can exist without a solid and secure core. Modern scientists proceed to investigate and neo-philosophers proceed to speculate on the nature of the Earth's interior, that it must be solid and well compacted under the tremendous pressure and gravity of the compacted mass of the whole planet, and mathematicians precede to make their computations, while volcanoes sporadically explode to let out heat and gas—caused by voids of entrapped by what, oxygen?

Not to demean academia, but knowledge today is censored from behind the scenes by murky paragons, no differently for the three branches of government, as all in common practice material "practicality", to the ruin of the people electing them. The Renaissance of 1453 is still far from completed.

Although Man intellectually accepts the idea, but only externally so, that the Earth is not the center of the universe, emotionally he counts on gravity, along with his earned surplus of paper money, to be his "practical" security anchor; and to him, gravity is pointing at his globe's centroid.

To Man, the Earth's centroid now has to be the epitome of massiveness and gravitational attraction, of solidity and ultimate material security (but not the ultimate Heavenly security), which is the new center of his scaled-down and faster-moving universe. Man securely lives high over the core of the Earth, whereupon he tries to build and control vain empires with sorts of Epicurean progress, all on borrowed fiat paper money. He pays his interest on fiat money with devices convertible to precious commodities and the land upon which he eats and sleeps.

Many of the newer religions of protest, misreading and misinterpreting the simple language of Holy Writ, help promote any kind of a physically secure Earth. A physically secure Earth is a dramatic alternate for destiny, until that time, when people hope they will not be here, when horrible happenings are about to come. While unfortunately, some of mankind has come to disdain his Earth, as though it is a mere expendable commodity, as though evil and inferior to his accumulating treasury of printed and recorded spiritless numbers ($) kept in strongboxes for when there is need for it, as if only that of all tangible and perishable instruments will help him survive, along with his yelling out, "I am saved!"

As Man matures and becomes more sophisticated, he speculates: Because pressure causes heat and heat causes melting, it is a given, therefore, that the core of the Earth must be composed of a superheated molten mass, sophisticatedly a near justifiable departure, a new compromise to solidity. Besides, it answers to his culture's hell, now that sophisticated Atheism among the "arrived classes," 1930 to 1989, has waned. And therefore, no need to investigate. Because the densest of elements are the metals, it is opined that the molten mass in the core has to be none other but of very hot and dense vapors of metallic origin, so that it yet may be characterized as solid. According to the conventional powers of imagination, because of the extreme of gravitational pull, even the superheated metallic vapors are compressed to a mass just as dense as cold iron, if not to a denser metal, but yet remaining hotter than ever. Well now, rationally, what is happening with hell? Where can it be centered under such conditions? (What lobby is at work here?)

Any form of heat in the Earth's core is not too difficult to understand, for it is in conformance with the belief that hell must be a domed abode in the Earth's center, based on some older Dantean belief that hell is down there, and that it is hot. Others envision a "core," cold and most solid, within the hypothetical molten magnum. Could these latter be a new breed of Atheists? Why can it not be of gold, to launch a new abysmal gold rush to save the debt or negatively based so-called "economy"? If so, the alleged Nephilim of the underworld must have already taken interest and possession thereof.

Jules Verne comes along happily to penetrate the Earth's deep solidity with intriguing caves interconnected with tunnels, most of which, of course, we know, just have to be flooded, for they are below sea level. Are these liquids within a temperature of our comfort zone? Is all this in a world with a cool or hot core? But Jules Verne wisely presents the venture as being within the standards of our human comfort zone.

We then enter the age when Man imagines levitating, or being levitated, as a common occurrence; that is, being lifted without force or propulsion but rather by magically floating upward. With propulsion, there are active dynamics that take place, which relate between Man and his solid Earth, while with levitation, the dynamics appear to be between Man and the Heavens, now securely alleviating him from any fear of falling. By resting on Heaven, or rigidly being suspended from there, a new sense of security is gained, not only "rationally", but emotionally and optimistically as well. For if the Earth can hang in the Heavens and not fall, so ought the individual person, independently, once prepared not to fall, in allusion to a New Age cult of supermen. Perhaps this broadened sense of adventurous security is based on his accepting the experience of space travel and return, to and from planets, whether bigger or smaller, firmer or looser than the Earth, and of course based on religious beliefs about the Heavens, thanks to the "missionary" work of *Star Wars, per* television, etcetera. Really, as we can see, the Heavens up there need not to be held up by pillars rising from the Earth.

The rapture doctrine of some religious sects may help increase credibility in a less solid, but allegedly a more evil Earth. As for Buck

Rogers, because the author thereof presents his story as fantasy with a gleam of unquenchable hope, he is not forgotten. In the meantime, in his tumultuous daily life, modern Man is losing his responsible control of his political, hygienic, and economic destiny, which causes him to lose faith as well in his Earth, be it solid or whatever, which now is becoming finite, crowded, and facing sorts of climaxes, with his allowing bankers, politicians, and judges to take everything away in their way of expressing thanks to the people, and unfortunately, with the people saying: "They can have it! We are gloriously going to be carried away, for we are saved, because we say we are." This is social irresponsibility. The ubiquitous social conspirators love these "holy rapturists" of passivity and irresponsibility playing into their game.

As Man is finishing the discovery of the entire surface of the Earth, he is thinking he could become the master of the Earth step by step, that he could be master of a delimited cosmos, but obviously, it would be only of a limited environment, lying within the scale of his self-proclaimed enforceable practicality. Not mankind in general, but rather it is some of the coy bankers, high clerics, politicians and judges, fancying, that by using Man they could become masters of this Earth step-by-step. Humble Man in his weaker "practicality" has come to accept such de facto frauds of his destiny, rather than to accept the Creator of him and of all that is.

Now, a century later, knowing he can travel safely beyond any of the former earthly limits, which he still really does not comprehend, Man is ripe to accept that the Earth itself is like a spaceship, not any longer a foundation upon some solid firmament below. He knows that he still does not know as much about the ocean floors as he does about the surface of the Moon, of Mars, of new galaxies, and he knows nothing about the inside of his Earth. He merely speculates along lines of conforming sophisticated fantasies, which result frivolously into attacking Mythology and Classic knowledge, which much of mankind, as self-styled modern pragmatists, will not and cannot understand. Mankind, in his thirst for friendship and love, in spite of losing faith in this world, wants to believe there are people on other planets, and possibly in the Earth's inner world, that will listen to him.

If true, and if it could be demonstrated that the Earth is hollow, Man and his scientists today are emotionally set to entertain the acceptance of such axiom. Psychologic mixed feelings, reinforced by current social turmoil, are giving way to the idea the Earth is not that secure, even in a physical sense.

Unfortunately, the youth sense inevitable doom from nuclear warfare, because of the alleged intransigence of selfish adults, which adults will not, cannot admit that they are beholden to secret paragons, especially paragons which had set them up as beholden "community leaders" over community conformists.

Man may be ashamed of not knowing his Earth, while already knowing something about his galaxy and other galaxies; and oddly, he still passes judgment on God and tries to correct his neighbor, while not knowing himself. To his credit, though, he is trying.

(The undersigned writer does believe joyfully that our Earth is secure as for its continuity, in spite of the dismal events of our times, albeit subject to periods of devastations due to reactivity against a paramount wrong thinking.)

The established scions of the new "scientific tradition", instead of the traditional "episteme", may still be the biggest obstacles to understanding a hollow Earth. Admittedly, however, the deliberative and deductive scientific methods are no longer subordinate to the experimental or empiric; to wit: The theories on the physical atomic structure and the theories on the activity of the atom have produced visible and tangible results according to expectation. So the dogmatic "Science a la Solvey" is obliged to do some back stepping.

Our Earth, with all sorts of problems, is in dire need of our love for her. With the right human vibrations from collectives of mankind, the Earth cannot but with favor respond. Loving the Earth bespeaks loving the Creator thereof above all else.

— ——— —

5 — UNEXPLAINED OBSERVATIONS

All scientific theories to date, even the farfetched, are secured on their having a focal point of gravity, allegedly fixed on the centroid of the Earth's sphere. But why does the force of gravity vary from the equator to the poles, and from over land to sea level, and over the sea at any given parallel? How should it be possible for anything on the Earth to escape gravity orderly, and to return just as orderly and safely?

The disappearance of the ocean currents into the poles is unexplained. The accumulation of volumes of water freezing at the poles, in spite of the poles being subjected to more stray reflective light, is not sufficiently offset with the comparatively few icebergs breaking away. Icebergs are of fresh water; they are not formed from ocean waters, not mixed. Could there be that much snowfall over the polar areas, to where actually fewer clouds and less precipitation may venture, because of the Earth's spinning them laterally away? Both polar ice caps are accumulations either of chance precipitation or of purer outflows from below, from the Earth's interior, counter-flowing next to the inflow of saline ocean waters but at separate hydrostatic zones.

The asymmetric misalignment of the magnetic axis, lying askew to the geographic axis and frequently shifting its location is left unexplained; as well as the source and cause of the aurora borealis and aurora australis. Could this be true of a hollow Sun as well?

When pressure causes heat, and when the imagined and accordingly alleged "core" of the Earth is hot, why do the oceans get colder and darker at their depths? Lack of sunlight is an insufficient answer. Spacecraft flying beyond our atmosphere and always being exposed closer to sunlight in space also experience extreme cold, except when going through the end zones of the atmospheric belts.

A terrestrial core of molten or vaporized matter under heat presupposes an eventual consummation. Mathematically, such consummation should have already consumed the Earth's alleged "core" if oxygen is present, of which plenty of supplies are needed to sustain any fiery heat in the core.

A solid Earth, with increasing cohesiveness toward its alleged core, would be contrary to certain known basics in the study of physics, especially in the "laws of spinning bodies". How can the Earth be singled out to be a colossal exception to the law of centrifugal forces? Accepting the concept of polar openings may help somewhat to understand the Earth as a form of kite, albeit there are other forces superseding aerodynamics, such as the lifting up against gravity of the life-sustaining electric eddy currents, etcetera. The Russian Navy now has torpedoes traveling underwater, not by forcing their way forth, but by creating a vacuum sucking them forth, keeping dry, at four hundred kilometers per hour.

How can the Earth, and other planets, floating in space, be solid when the fowl of the air, kites, and airplane wings work only because of some aerodynamic hollowing out, associated with the flowing of negative air pressures, in their configuration? How can the Earth be a gross exception?

How can any of the heavenly bodies stay suspended in the Heavens in accordance to some definite and orderly government, if each does not actively contribute to the forces interfacing and interacting with all the other bodies? Then why do we say planetary alignments influence our Earth, as if such are the only influence?

It is simple to say abstractly that the Earth's globular shape is molded by the balancing of centrifugal and centripetal forces, but what causes these to have an allotted location in the galaxy, a source, a footing, the knowledge to be a sphere and to be exactly there? Do these forces have other matrices, whereby, perhaps, the molded shape of the Earth is the summation of strange vectors crossing and defining a shell?

Would not the celestial government of planets, stars, and all other kinds of bodies and clouds as well have to be based on some distributive aerodynamic-like principle and any kind of kite-like or other specific configuration?

When one of the heavenly bodies fails, why doesn't a chain reaction of overpowering attractions, or oppositely of repulsions, take place, steering into a near cosmic collapse – an implosion or explosion, or a domino-like orderly collapse? But we do speculate on planet alignments influencing our Earth.

According to ancient tradition, the Earth is the fourth planet outside the Sun, but today, objectively, it is the third, Venus being the second. Ancient tradition holds that an asteroid system orbiting between Venus and Earth is the remnants of a primordial planet.

Planetary and astral attraction-repulsion, then, must be conditional and not innate; each must be more than simply a balancing chip, a shim, contributing to a check and balance of a greater whole. It must be alive with incomprehensible energies. As for celestial "accidents" or random happenings, per some sophists, such as in the emergence of comets, in reality, can there be a simple form of negative balancing to maintain a better balanced as-of-yet greater whole?

But for planets disastrously to be crashing and disappearing is the epitome of solid Earth thinking. Did such ever happen? The magnetic, atmospheric, and series of exospheric and atmospheric spheres of the planets would cushion and repulse each other before contact.

Why should all the celestial bodies be rotating in the same direction, ignoring the need of certain opposite rotations, as in a gear complex, in order to maintain a balanced order? Do planets turn one way and stars the other way? Are planets hard rocks and stars burning gasses, as conventional references want these to be?

Why has not the Earth, nor the Sun, nor other heavenly bodies, been dissipating any of its momentum in spinning throughout the millennia? What rejuvenates them? Why do comets keep burning out, but do not seem to finish burning out? Are they being fed fuel to keep the exosphere supplied with a degree of alkalinity?

Is it possible for a deep-rooted exploding volcano to split the Earth into two or three fragments, and not allow a reconstituting of these into a whole Earth? Could each fragment seek another government and direction, with a separate reshaping? Could each break loose from the given magnetic and electric forces of cohesion? Under what laws could each restructure itself spontaneously? Or magnetically, or centripetally, and reconstitute itself in the original whole Earth? Could this phenomenon be related as well to comets or stars?

Why does the Sun "endlessly keep on-and-on burning out" as is often said, but it is never done doing so? If allegedly half burnt out, why have the celestial balancing forces not lost, compensated, or altered their sizes and natural government? Perhaps some of these did happen, and here we still are.

Why does the rule of "equal and opposite reaction" still work in the near infinite void of zero pressure of outer space when that rule rationally is related to sensual or tactual physical forces? Is it an understood complement, that the passive part of the "equal and opposite reaction", of something non-tangential, too, is valid? "Do unto others as you would have them do unto you." Is that not the same law of "equal and opposite reactions"?

Why do progressively fewer and fewer leap years occur through the documented span of two thousand years, per the equinoctial (Gregorian) calendar?

— ——— —

ON A SPHERE

Laboratory-simulated tests enforce certain logical precepts on the nature and defensive quality of our Earth's structure.

6 — PROCEDURE INTO THE MATTER

The Heavens are of a divine symphony, a celestial dance, an epitome of orderly and compensatory rhythmic motion. The ancient Egyptians and Pythagoras of Croton found that all heavenly rhythms, distances, circuitries, and trajectories conform to the two forms of the golden mean, the 0.618,033 and the 1.618,033, each of which divides into unity to yield the other. The multiples and powers of these two forms of the golden mean fit into most celestial and planetary measures. The Orphic poems demonstrate there is nothing random about the heavenly dimensions and motions. So much more is unexplainable, but suffice it for others to begin investigating somewhere.

The divine symphony of the Heavens has its image in the symphony of the world to which we are sensitive and to the symphony in the mini world to which we may scientifically speculate. We may advance even yet to a finer world beyond instrumental possibilities, so petite that it yet defies any kind of scientific probing, other than dialectic speculation.

– ––– –

A theoretic case of a ball at hand, or hollow sphere, is scrutinized and portrayed in the confines of a containing mold, in an experimental laboratory. Then, paralleling, follows the development of the very similar case of our planet Earth being formed, also a sphere, being contained in the atmospheric and celestial mold formed of other kinds of ethereal and celestial forces.

The difference in the two conditions is: The apparatus in the laboratory is assumed to be fixed, as to the floor; while our planet is figuratively in "suspension", floating, but with its orbital plane serving as a celestial floor. The difference conceptually may be mitigated when, for instance, a sphere or balloon, is filled tightly with a gas, which is lighter than the air we breathe, or is charged with, say, antigravity magnetism. So, if the containing sphere is not heavily secured, it may blast through the ceiling and away, not by outside propulsion but voluntarily, through its accumulated energy.

However, the objects spaced out through the Heavens are seen by us mortals as randomly floating or in random suspension, randomly tacked onto the celestial dome. Perhaps it is of a very sound network of electric, magnetic, and other means, and perhaps it forms a single organizational web-like essence. In illustration, milk is in several bottles in cases, the cases are stacked upon each other, while the milk from milk is in unquestionably fixed isolation, but it is yet collectively as one. It is milk. Eventually, the scientific "knowns" in our mini universe and the celestial "unknowns" in our maxi universe may telescopically converge to a single fact, a single truth.

A sphere is defined as being round, and it can roll any which way on a smooth floor. A ball bearing, a beach ball, or a billiard ball ideally may approach sphericity. However, the Earth is not a perfect sphere.

When dealing with a "sphere", which betrays an axis of rotation, wheel-like, it may not roll any which way, but it may turn in place or roll in a given direction. When looking at a sphere in section, cutting across the axis of rotation perpendicularly, it displays a circular symmetry. But when cut collinearly along the axis of rotation, it may not display circularity, speaking in subtleties, no more than does half of an apple cut through its two navels. Nevertheless, in both cases herein, we call these "spheres", but we must keep the distinctions in mind.

A perfect sphere of uniform consistency and form with an imaginary centerline may be taken as dead or passive, and as quite immune from reacting internally to thrusts and vibrations. An imperfect "sphere", however, can become alive as a capacitor of energies or a conductor of energies. Accordingly, these energies may further deform an imperfect sphere, as nature finds necessary. A billiard ball does exactly what you would have it do because it is non-programmable and with respect to symmetry in any which way it is dead.

The initial procedure herein is: First, the laboratory "theory" is given. Second, a sketch elucidating the concept follows. And third, there is the "earth-wise", or terrestrial-wise, rephrasing of the theory to the sketch above.

— ——— —

7 — CERTAIN KNOWN FACTS

From a look at physics, certain laws are readily accepted as appropriate for most any immediate subject of studies. Whether through some caution or timidity, wisely or unwisely, however, such laws are hardly applied, or totally ignored, both in the more universal or Heavenly studies and in the more specific accumulated data about our planet.

Known: A mass, whether solid, spongy, with hollow center, or of whatever shape contained in a globular mold, while spinning about a spindle by centrifugal force, tends to distribute and impact most of its mass on to the outer walls of the mold along its circumferential peripheries, furthest away from the spindle (axis) of rotation. The denser granules and-or more viscous liquids easily tend to pile up at the farthest limits, the circumference, of any irregularly shaped spinning mold, and the lighter granules and liquids lag and pile up last to form the inner surface of the rotating mass. A tensile thinning, or disintegration, or a vacuum begins at the core of the mass as it is set into spinning, working outward, especially away from along the axial line defined by the upper and lower spindles. Plain water in a spinning globular container will centrifugally push the water outwards and accumulate a huge air bubble along the line of spindle. Hence, there is reason to begin suspecting that all twirling masses have the propensity of hollowing out.

Earth-wise: Translating this into terms, theoretically, for the spheric Earth: The physical matter of which the Earth is constituted, because of its revolving about an axis of rotation, while being contained by its outer surface, which by whatsoever kind of forces assumes the form of a sphere, impacts its mass toward the delineated peripheral limits by centrifugal force. The centrifugally guided impaction is greatest in amount and is distributed at the farthest outreaches, from the axis of rotation, which is at the Earth's circumference, the equator.

— —— —

Known: If the whirling mass in a globular mold is quite rigid or solid, there will be crushing and greater compaction of matter toward the peripheral walls. And a tensile thinning, disintegration, begins from the core and works outward from the imaginary spindle, leaving an internal vacuum. Unless air seeps into the hollow, the vacuum may subject the configuration to imploding.

Earth-wise: There is reason, therefore, to begin suspecting that the Earth is hollow, or at least has naturally accrued a hollow, and the thinnest part of the walls is adjacent to the poles. Air seeps into the hollow to bring the atmospheric pressure within and without closer to equilibrium. Outside air seeps into the hollow from wherever there is least resistance, which generally is at the poles. Pressure causes heat, but a vacuum and lack of friction induce a drop in temperature. Air seepage inward, or hydrostatic action, forces the lightweight air bubbles inward to the Earth's centroid, away from the outwardly concentrating mass. The centroidal environment is subjected to rarefaction; hence it ought to be cold, or cool.

— —— —

Known: A whirling spheric mold, when its rate of revolving is being lowered, the stability of its assumed centroidal line between the spindle points tends to pull the spindle rhythmically sideways, and it begins to waver to absorb its losing inertia, while readjusting through a series of compensational shifts. Wobbling takes place in the migration and change of a centroidal redistribution of the mass. And the slant of its line of spinning, relative to the floor, increases. Through a series of diminishing overcompensating shifts it differentiates and threatens distortion. The globular mold's shape must be of plastic quality and variable, as it is subjected to a rising temperature through induced friction, and to resist permanent deformation as it is twirling.

Temperature changes reflect into a change of shape, as well, due to the dispensation of heat and friction.

Earth-wise: When subjected to lowering its rate of revolutions in a given time frame, the Earth's axis of rotation loses its assumed verticality, giving into wobbling from perpendicularity, to absorb the access or diminishing momentum, which yields into a vacillating modicum to prevent any traumatic shock. Hence, the Earth's shape must be plastic and variable, while being subjected to a rise of temporary heat, to resist irretrievable loss of its spheric contour and disintegration from failing to dissipate its momentum, or to turn into a halo of dust. Also, freeing itself of any fixed celestial spindle, its internal activity motivates it to meander and seek compensation while orbiting around its original heavenly fixed spindle location.

— ——— —

Known: On any whirling mold, spheric or irregular, temperature changes reflect into dimensional changes, due to the dissipation of heat and friction. Warping therefore occurs temporarily or permanently from the irregularity of dissipating heat with the resulting expansion of molecular activity.

Earth-wise: The temperature changes distort its consistency into endless configurations because of the rearrangements of the various materials of which it is formed. Temperature changes differentially befall certain parts, or the entire surface, or the whole of the mass. From night to day, from season to season, era to era, from volcanoes to popping up terrestrial plates, besides other factors so contributing, the Earth never is the same from day to day.

— ——— —

Known: The wobbling is compounded in any new irregular distribution as more heat builds up from the greater pressure of crushing in the revised configuration of the mass, due to the impending relocation and changing slant of its imaginary spindle line of whirling. Wobbling is bound to occur simply in the slowing speed of whirling to dissipate energy, and not always without leaving some permanent rearrangement of matter. Wobbling increases the friction with the air all about.

Earth-wise: Terrestrial wobbling, resulting from the age-to-age changes of average temperature and the friction from cyclically undulating, consequently forms and reforms the layered stratifications of continental plates.

— —— —

Known: A gyroscopic governor, set to spin on its momentum, increases its rate of rotation when its arms drop to maintain its equivalent momentum, and vice versa, and many times over again, without altering the rate of dissipation of energy. This is true also of a spinning sphere; it slows when widening and speeds when slimming. When forced from its strictly vertical or horizontal orientation, a gyroscope wobbles in compensation until its restoration to vertical or horizontal normalcy is reclaimed. In the meantime, a potential of energies is stored. It can be expected in the meantime, that derivatively there are subtle interplays between momentums and inertias.

Earth-wise: The Earth's twenty-four-hour period cannot be an absolute measure, but it is variable from time to time. The rotation of the Earth about its axis declares the standard for the length of time of the twenty-four-hour period. However, the rate of rotation is variable, as the Earth's diameter keeps changing. And this is a challenge to Man's calendric accuracies. A full orbital circuit may contain greater or fewer amounts of days, between 360 to 370 days, and its completed orbital cycle may be faster or slower, independent or not of the number of days. The intensity and frequency of solar and other rays are also altered or refracted into an infinite variety of forms of energy upon resistance to any substance or condition.

— —— —

Known theory: By Kirchhoff's theory, when heat is applied to any point on a heat-conducting body of any shape (rod, pipe, plate, irregular casting), ions move toward the heated portion, causing a drop in temperature at the farthest points away from the applied concentration of the heat. The condition lasts until the spread of the heat engulfs the entire body. Also observed: when the application of heat is removed, but before the heat is uniformly spread throughout the object, the ions quickly reverse to assume their former positions; and with the ions the already stored and concentrated heat spreads more quickly to the yet unaffected farthest areas, to be distributed evenly at the compromised temperature throughout the whole body before cooling sets more uniformly throughout the mass of the body; but with the cooling naturally spreading more quickly to the edges and surfaces, due to heat loss principles. (See following figures.)

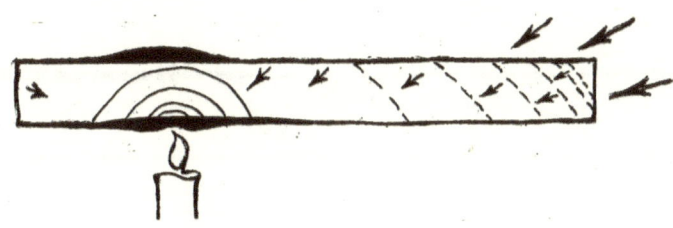

7.1 — Ions from farthest areas rush toward the applied heat.

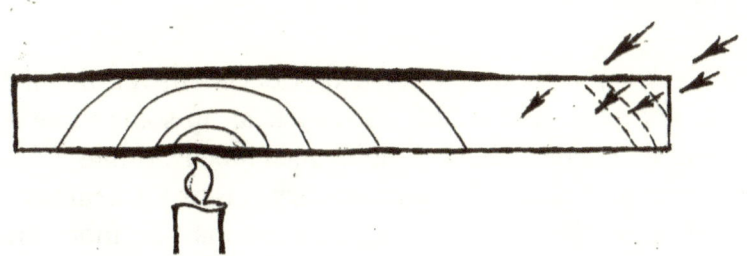

7.2 — Temperature drops in the wake of travel of ions.

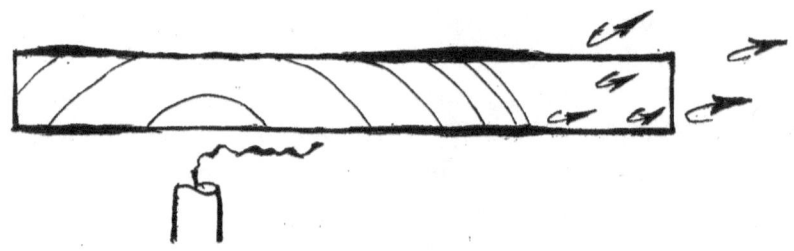

7.3 — Heat cut off, ions return, as the heat uniformly spreads.

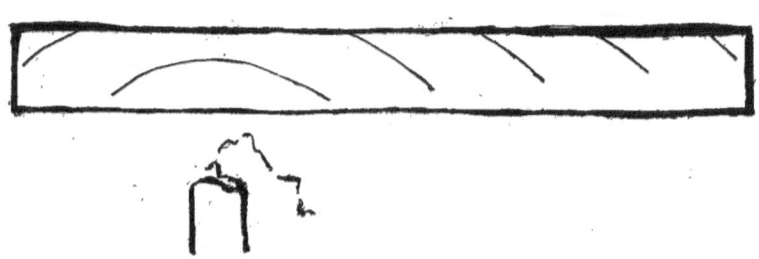

7.4 — Retained heat automatically is distributed throughout evenly.

This phenomenon is often experienced in the metal handles of pots and pans in cooking, as they heat up some more after the source of heat is turned off. Complicated metal castings are designed sometimes with seemingly unnecessary fins or thickenings of parts, but this is so that their cooling process may be more uniformly controlled to prevent fractures.

Earth-wise: Assume the bars in the above figures are crescents, or longitudinal segments dividing the Earth through the North and South Poles (like dividing a cantaloupe). Both polar regions of the Earth become cold, not only for the lack of intense perpendicular sunlight, but more so for the heat at the equatorial belt drawing ions from the extremes of the Earth. Heat builds up internally, strongest just under the equator. Centrifugal and centripetal forces working against each other exert their greatest pressure at the equatorial regions, and they lessen toward the poles toward naught.

— ———— —

8 — WHIRLING SPHERIC MOLD

On proceeding, bear in mind the possibility that a celestial government, not unlike the given spheric or globular molds in a laboratory, may have formed the Earth according to an ever-repeating pattern, described as follows.

Theory: A physical elastic container shaped like a globe is whirling smoothly about its line of spindle at a moderate speed as some liquid or non-viscous pliable mass is being injected into it through an opening, the navel at its upper spindle.

The whirling watery mass distributes itself on the inner surface of the mold, thickest along the farthest reaches from the spindle line, which is under the circumferential region, because the greatest centrifugal force occurs at the circumference. The mass diminishes laterally toward the spindles, because of the diminishing centrifugal force, to thinnest, approaching zero. The spindle points in accordance with the known laws of physics yield a navel at each end (akin to most any fruit).

The inside surface of the whirling watery mass of the spheric globe assumes the internal shape of a lemon (a prolate), whirling about its stem-to-root line. In section, the inside is like an ellipsoid coming to tangency at the spindle points with the spheric mold at the spindle line, defining the two nodal points, or navels.

In cross section, the globe appears as two crescents opposing each other, but touching at their tips at their common upper and lower navels. That is, the inner and outer surfaces come to tangency; the inner surface forms as a spinning prolate, while the outer surface maintains its sphericity. Or said otherwise, in section it is a vertical ellipse revolving about its long axis, within the revolving near-perfect spherical globe, and likely at a different speed of spinning. (See figure 8.1).

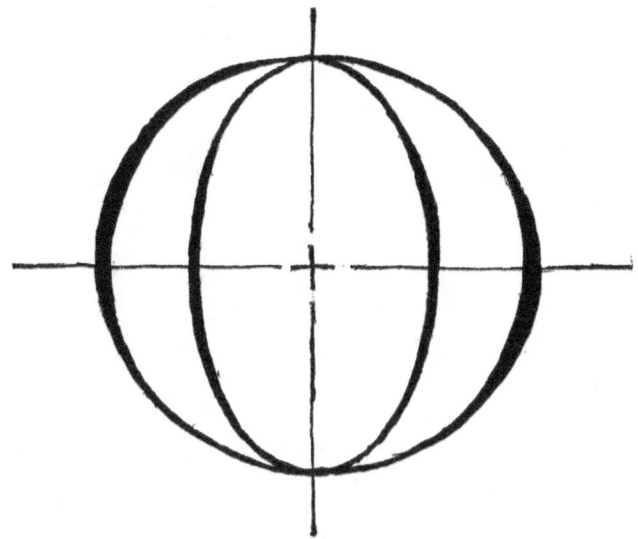

8.1 — Centrifugal force distributes lose mass contained in a spinning globular container, thickest at its equator and thins toward both ends.

Earth-wise, primordially: The outer surface of the Earth, in its initial coming into being, may have been a near-perfect sphere, and the inner surface a prolate with its long axis sharing the same axis of rotation of the outer surface. The chief substance of the Earth is watery. Although the rotational speeds of the outer and inner surfaces may not be the same, while slipping between each other, there is hardly any friction, hardly any resistance, from the stirring watery mass, except for the simple centrifugal force of rotation about its axis.

— ————— —

Theory: Two infinitesimal holes at the thinnest part of the pliable mass are formed at the coincident nodal points of the outer sphere (spheric surface) and the inner prolate surface about the line of spindle. In reality, the holes, or navels, are of slightly increased size to satisfy the cohesive properties of the viscous mass being turned inside the globular mold. Thus the edges of the viscous mass taking shape around the spindle line is rounded off and yields to other adhesive forces to hold its shape. (See figure 8.2.)

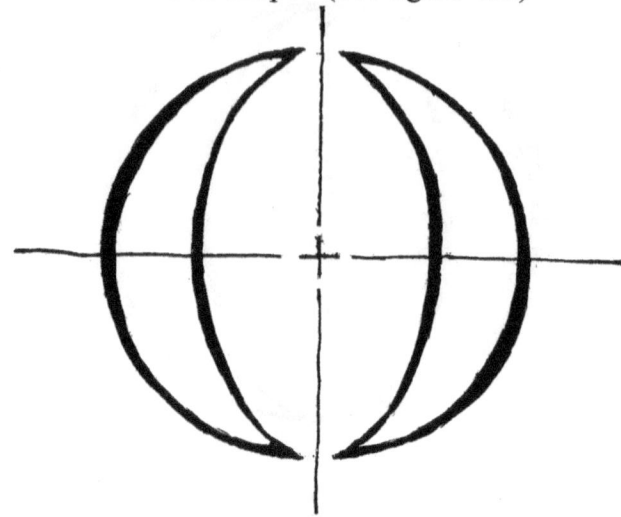

8.2 — The mass contained in a spinning global container is distributed by centrifugal force, as two navels smoothly open up on the centerline.

Earth-wise, hypothetically: The mass of the Earth rotating about its axis is pressed outward by centrifugal force from the axis of rotation, as the mass is distributed toward the Earth's outer surface. This causes the greater concentration of the mass to form under the region of the equator, gradually thinning to near naught at the oculi, at the North and South Poles on the Earth's axis of rotation. The limits of the entire earthly sphere are defined by the Earth's outer atmosphere, an invisible container, governed by the centrifugal and centripetal forces, among other forces, as well by the inner atmosphere, all coming into balance.

— ——— —

Theory: The accelerating of the rotational speed forcefully distorts the mold. It becomes a whirling tangerine shaped mold (oblate spheroid mold). The spinning circumference answers to the increased centrifugal forces and the weakened centripetal forces at the mold by spreading wider. The two spindle points, being the nodal points, are significantly less subjected to the centrifugal forces and pull closer together (by gyroscopic principle) along their spindle line. The free inside surface of the viscous mass facing the axis, also spreads toward its circumference, away from the axis, and in turn approaches the form of a sphere inside the hollow void, or hollow sphere. The aggregate volume of the mass keeps to a constant, but the outer and inner surface areas per solid geometric reason increase. The spindle common points of the exterior globe and the interior globular surface maintain their fixed tangency, on their common spindle line, theoretically at zero thickness. (See figure 8.3.)

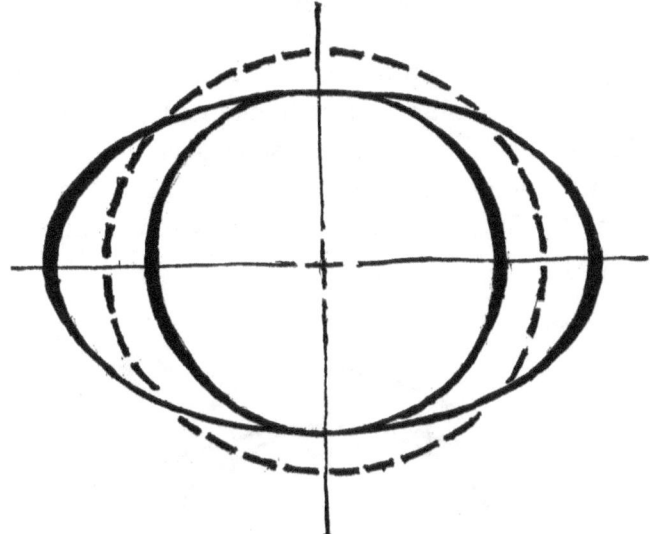

8.3 — The faster spinning ellipsoidal mold distorts the inner surface of the spinning pliable mass to approach internal sphericity.

Earth-wise, hypothetically: When the number of revolutions of the Earth about its axis increases within a given time, centrifugal force tends to reshape the Earth's outer sphericity to an oblate, increasing the circumferential length of its equator and pulling the North and South Poles closer to each other. Both inner and outer spheroids containing the Earth's mass maintain commonality at the Earth's two theoretic polar points.

Some older references claim that the North Pole is "flattened," which bespeaks an admission that the view is of the space of the polar opening, whether exposed, or covered over (or as with a membrane stretched across the opening like on a drum).

— ———— —

Theory: With lesser rotational speed, while the moment of inertia remains a constant, the eccentricity of the oblate globular container mold increase. The circumferential measure accordingly increases, and the inside surface of the watery mass also evolves into an oblate spheroid. Or, in section an inner ellipsoid evolves, but to less eccentricity than the outside elliptic configuration. The given constant of the volume of the contained mass is maintained. The spindle nodes of the two oblate spheroids, while remaining in tangency at their common spindle line of twirling, pull more so closer together along the spindle line to a more extreme tangerine shape. (See figure 8.4.)

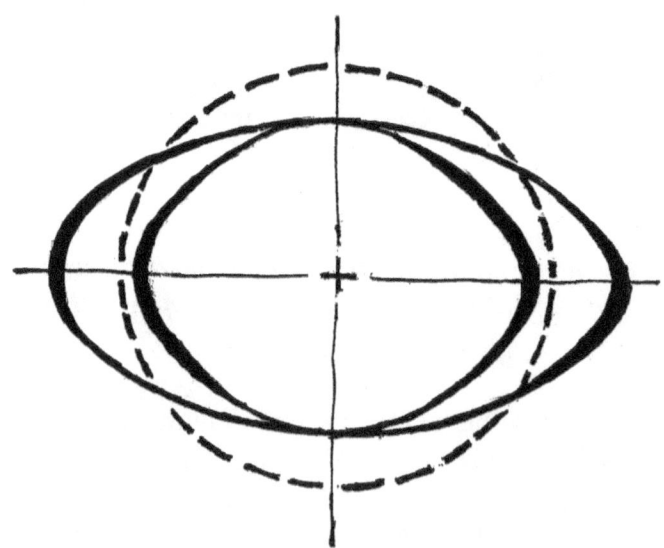

8.4 — Both the inner to a lesser degree and the outer surface to a greater degree become ellipsoidal in a decelerating spinning mold.

Earth-wise, hypothetically: When the Earth, during some other past zodiacal age, rotates at a lesser speed about its axis, the circumferential length of the equator is increased, while the distance between the North and South Poles is decreased. In addition, the theoretic equator of the inner surface increases proportionately, while the terrestrial mass theoretically remains a constant.

— ——— —

Theory: The watery contents in the globular mold orderly break open at the tangencies of the two spindles, due to their thinness and greater cohesiveness than adhesiveness, yielding two definitive navel openings, facing each other.

Aerodynamically, the oblate spheroid theoretically could obtain the capacity to be buoyed as a flying disc in space (see figure 8.5).

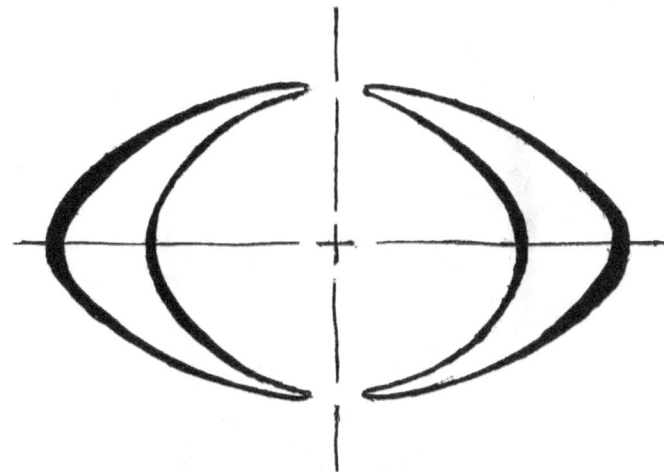

8.5 — A further ellipsoidal distortion of a spinning mold, on inner and outer ellipsoidal surfaces, the thinning mass opens the axial points.

Earth-wise: Due to their thinness, the inner and outer surfaces of the oblated Earth break open widely and orderly at their polar tangencies, yielding a definitive polar oculus at each pole. Aerodynamically, the oblate spheroid assumes the capacity of floating in space while in orbit.

The edges of the naturally occurring polar openings in scale are sufficiently rounded for a ship unwittingly to cruise around the lip of a polar opening into the inner surface, almost without noticing its departure from the outer world —as has been said— and to cruise back out, as if exiting from a dream world.

— ——— —

Theory: An elastic globular mold whirls its contained watery mass at a higher speed. The mass accordingly tends to distribute itself more evenly. The whirling elastic globular mold stretches in upright direction and shrinks around its circumference. (See figure 8.6.)

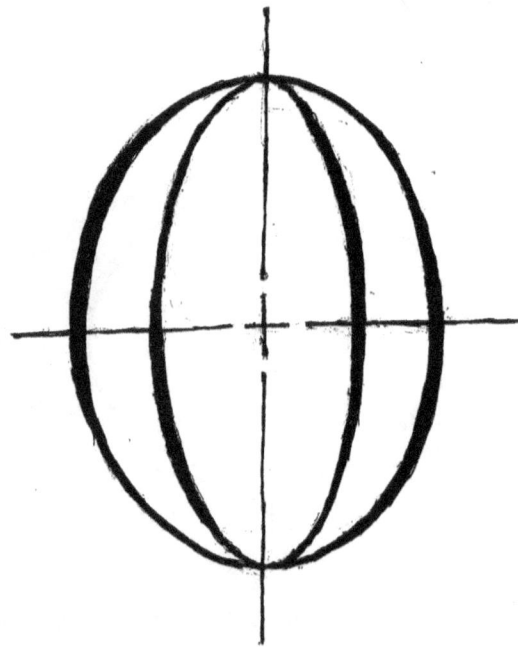

8.6 — A prolate sphere with its watery mass whirls along the inside wall.

Earth-wise: Although it appears to be possible, there is no hint of the Earth ever happening to become a prolate; theoretically it may happen.

— ———— —

An ideal primordial configuration of the Earth would perhaps be structured above and below a more perfect theoretic centroidal median sphere —between a slightly tangerine outer configuration and a slightly lemon inner configuration— the centroidal theoretic median sphere being the more perfect of the three spheres, where the centrifugal and centripetal forces would be more harmoniously equalized.

— ———— —

Theory: Both the cohesive and adhesive forces within the viscous mass, when the whirling about its precessional axis, whether rhythmically or erratically, the contents produce a reactionary vibration to absorb the greater complex of momentums. With its viscosity and tendency toward stratification within the mixture, the inner oblate spheroid is further deformed. When the exterior is contained in a stiffer mold, due to the understood forces, the interior ellipsoid absorbs most all of the difference from the transverse undulation, while striving for gyroscopic balance, and results into some deformation. The denser particles collect to form into a belt under the whirling circumference. The interior ellipsoid is distorted into a shape approaching a cylinder, centered collinearly on the line of the spindle, while in balance straddling the inner circumference. The lighter particles, which are more pliable, being shaken aside residually, tend to separate and to hold to the original globular form in the areas closer to the spindle areas. The resulting interior shape is from elliptic to a barrel form with dome-like ends, their two dome-like peaks being centered about the two navels of the spindle points; it is somewhat mindful of the aerodynamics of a "flying saucer", or UFO fiction. (See figure 8.7.)

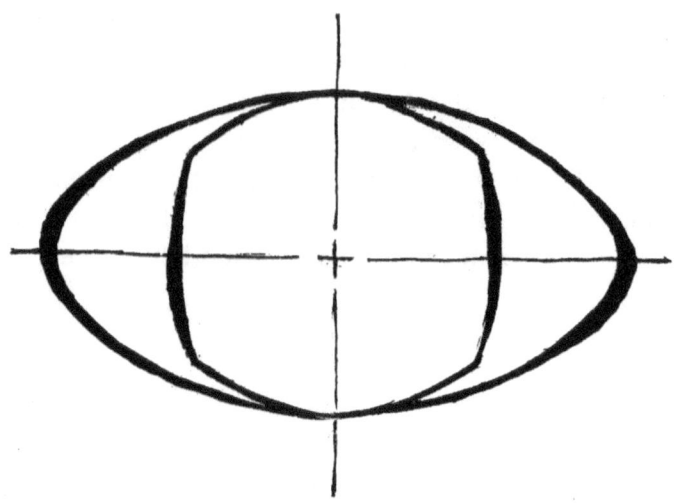

8.7 — Erratic rotational speed of mold increases spheric eccentricity and distorts the inner surface with differentiated stratifying contents.

Earth-wise: Within the somewhat pliable terrestrial mass, while subjected to a state of erratic rotational speed, accelerating and decelerating, with the dissipation of energy reflecting in some transverse sine wave vibration, and with plates of hard and softer matter being tossed against the equatorial zone irregularly, the Earth takes form with an irregular interior configuration. The looser matter finds its path leading away from the equatorial zone to concentrate toward either of the sub-polar regions.   The harder matter and terrestrial plates concentrate and stratify into layers, and come to a near cylindrical form straddling the inner equator; but the polar regions remain almost immune to any stratification or change.

— ⸺ —

Theory: The thin mass of the shell subjected to wobbling next to the navels, that is, next to the points of tangency of both outside and inside configurations near the flatness of the navel regions, is centrifugally tossed laterally widening the polar openings.   This allows the opposing two openings above and below to fluctuate, enlarge, and shrink independently of each other according to the differentiating natural undulating conditions, while straddling the line of spindle. (See figure 8.8.)

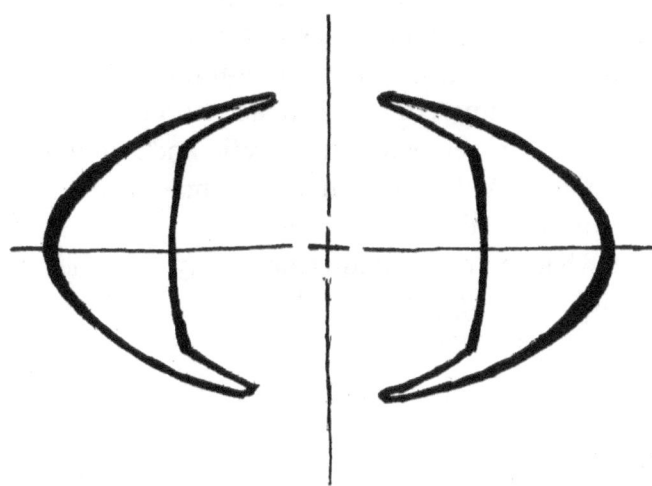

8.8 — *The polar openings widen as the thin mass edging at the poles dissipates centrifugally into the shell's mass, as to form new satellites.*

Earth-wise: The thin terrestrial mass fluctuating unusually more intensely about the apparently flattened polar oculi, causes the oculi centrifugally to widen. Aerodynamically the Earth assumes airborne qualities for sailing through variable weather conditions in situ or off and away.

The fluctuating polar oculi are mindful of ancient seagoing craft with ingeniously designed flexible sides cushioning and deflecting powerful waves without disturbing the safety and speed of the craft. Also, hot air balloons control their altitude by controlling the temperature of the air within the balloon, as by releasing hot air to lower its elevation and by heating the contained air to rise.

— ——— —

Theory: When the spindle points pull away from each other, the circumference of the mold shrinks, and the tangerine shape reverses to a sphere, and on to approaching a lemon shape. And the rate of rotating automatically increases, per gyroscopic principle, that is, with the energy potential remaining the same.

The global mold goes into a spheric, and to a prolate mold, or vice versa. These reversing effects also cause changes in the stratification and redistribution of the viscous mass within the ellipsoidal. They also cause and-or react to a heat and cold non-uniform differentiation.

Earth-wise: The immediate sub-polar regions of the Earth —that is, approximating the regions in the Arctic and Antarctic Circles— being more porous and lighter in weight matter, of thinner depth (as the thinning toward the ends of a crescent) to the opening, and having somewhat more fluctuating flexibility, are able to dissipate heat more quickly and with less thermodynamic shock with the Earth's atmosphere.

— ——— —

AN ERRANT SPHERE

A sphere, while in its own motion, is rutinely distracted from its particular location, affecting it internally and externally.

9 — WOBBLING SPHERIC MOLD

Theory: Any sudden wobble in a rotating spheric or oblate globular mold tends to toss the contained mass more violently away from the line of spindle and its proximities, and to bounce the mass, from above and below, into the circumferential region to begin the process of dissipating the accrued potential of moment. The wobbling increases the size of the two navel openings by laterally evidencing some turbulence.

More of the mass is tossed toward a widened circumferential belt, with its width defined by the upper and lower limits of wobble—not toward the actual circumferential plane of the spinning mass. A horizontal gyroscopic plane tangentially passes through the circumference of each navel opening, presenting a cyclical internal wobbly distortion with each rotational cycle, while bouncing above and below the orbital path (as with a twirling coin tossed on a table).

The mass underneath straddling the circumference of the mold absorbs the wobble, while its mass swells inward into the inner space, concentrating its mass to a smaller inside radius relative to the spindle, and the vertical distance between the navels is drawn shorter. The altered inner space comes to a pinched-in waist or a spinning figure eight in section; the section of the shell is like a bow weapon with a thickened handgrip at its center. (See figure 9.1.)

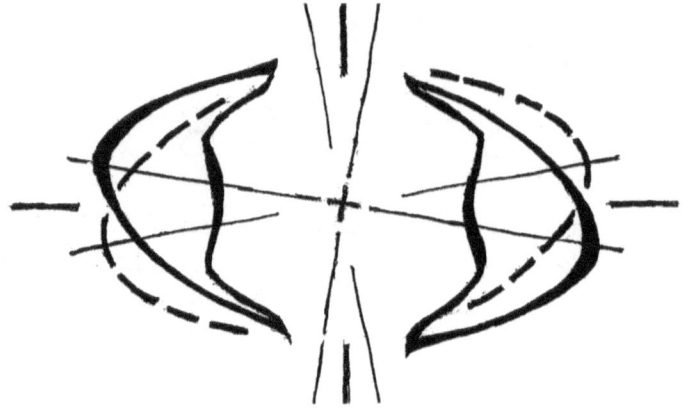

9.1 — A mold spins, wobbles and tosses the contained mass into a widening and massively swelling circumferential zone.

Earth-wise: As the Earth becomes an oblated spheroid, its rotating speed reduces as it yields to some wobbling to absorb the excess energy until it re-stabilizes. The internal mass in incremental stages tends to straddle the two diverging spaces between the equatorial plane and orbital plane —as their intersection shifts from end to end of the orbital ellipse— forming an inner ridge ring, and swells into the Earth's central concavity. The distance between the inner equator and the axis of rotation is reduced as it bulges into an inner ridge, straddling a new mean equatorial plane between the extremes of the wobbling above and below the orbital plane. The inner ridge ring would be somewhat more pliable and of lighter terrestrial matter.

— ——— —

Theory: As the rotational speed of the mold is decelerating, a stronger wobbling of the mold causes the navel edges to recede axially inward, becoming dimples. The oblate spheroid mold assumes an outside form closer to a wide round apple, with the two navels lying in the recesses. The dimpling effect stabilizes the width of the navel openings, as the diameter of the openings must keep under the distance between the wobbling line of spindle and the lips of the spindle openings. The turned-in lips focus conically upon the centroid of the body. (See figure 9.2, arrows show minimum-maximum radius of openings.)

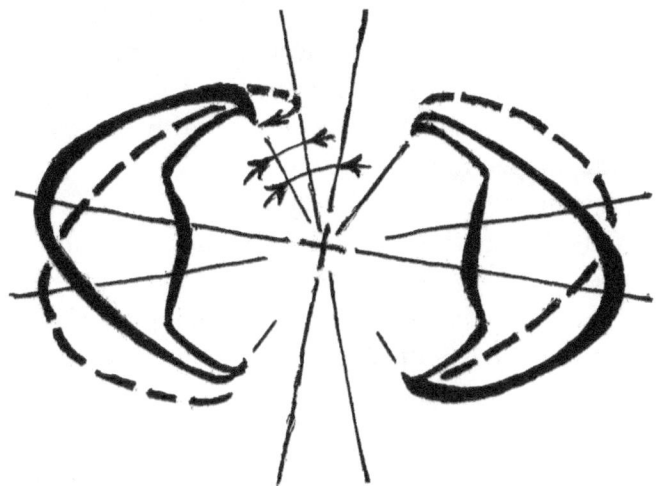

9.2 — Edges of navel openings subjected to least centrifugal force, while wobbling recede into the spheroid and point to sphere's centroid.

Earth-wise: The wobbling from the dissipating energy causes the polar edges of the terrestrial mass to recede inward, pointing toward the Earth's center. The Earth's wobbling from the dissipating energy, as the rotational speed of the Earth is decelerating, draws the thinned terrestrial mass at the polar edges to tighten toward the axis of rotation. The width of the polar oculi is confined between the nearest distance of the wobbling axis of rotation and the edge of each of the oculi. Accordingly the ocular lips turn inward pointing to the Earth's centroid, as two conics are formed with apexes at the centroid and bases at the polar openings.

— ——— —

Theory: Assume the mold is settled into its new shape, the wobbling has nearly disappeared, and that it is somewhat elastic but with a rigid circumference straining to keep true to the horizontal gyroscopic plane. Its line of spindle leans off verticality (or, say, the table is tilted, while the line of spindle maintains verticality). A rotating mass naturally does not deviate from horizontality, or from verticality. Otherwise, it causes strains, friction, and distortion of the mold. A cyclical distortional wobble is induced as conditions and distances cyclically vary between the circumference and the theoretic tilted spindle nodes. The condition becomes more acute for a tangerine-shaped mold.

The immediate solid areas next to the spindle nodes, having the lighter mass and fluctuating, tend to keep truer to the slanted spindle line. The navel openings tend to hold to parallelity with the horizontal gyroscopic plane. While wobbling to maintain horizontality, the positions next to the navel openings, cyclically become distorted.

Each revolution yields an obliquely disturbing wobble due to the angle of the spindle line. That is, sections of the equator, midpoint areas between the circumferential area and the spindle areas, cyclically stretch and shrink, while struggling to keep to horizontality. Said otherwise, with each rotational cycle, the distance between the opening lips and their circumference diagonally expands and contracts, alternating between above and below with each half cycle, straining and distorting the limits of the terrestrial mass.

This distortional trapezoidal dynamism could work only when the proper rhythmic undulating timing is attained, in order to prevent destruction. (See figure 9.3, arrows indicate belt areas of greatest gyroscopic disturbance).

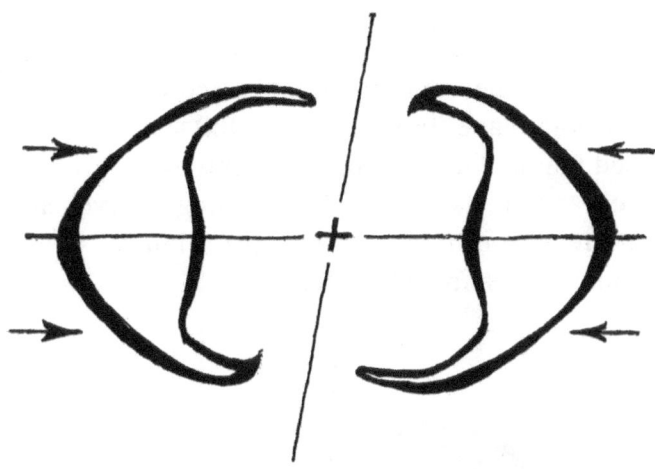

9.3 — Circumferential mass elastically adheres to gyroscopic plane in spite of angle of the compensatory wobbling centerline of rotation.

Earth-wise: Assume the Earth is stabilized in its new oblate form, with a fixed equatorial magnitude and with its polar oculi set parallel with the equatorial plane, while the axis of rotation is askew to the equatorial plane and to the two ocular planes. Of necessity, the shell of the Earth is subjected to diagonal and torsional strains, the greatest being in the two zones each between the tropic circles and the polar circles. Strains are mitigated by the varying coefficients of compression and tension, facilitated with temperature changes and forms of resiliency. The equatorial plane and the polar ocular planes wobble slightly in compromise to mitigate reactions, until reaching a tolerant undulating timing.

— ——— —

Theory: Cyclical distortion, receiving a thrust with each revolution, can become rhythmically stabilized with a stabilized rate of rotations and a certain established consistent undulating elasticity of the mass, a new complex rhythm.

Some particles at the edges by the openings, rather than being tossed laterally by the wobbling, may tear themselves off from the mass. Such may be caught into islands or clouds of isolated matter, trapped, and kept suspended and bouncing within the space of the navel openings. These islands may keep clear of the wobbling spindle line as they seek a new natural spindle line of their own vertical to the gyroscopic plane, breaking from the slanted spindle line of the contained mass. There is some discontinuous grinding friction between the trapped islands at either opening, as cyclically there is pressure and release in their contact. (See figure 9.4, arc between spindle opening and spindle line noted.)

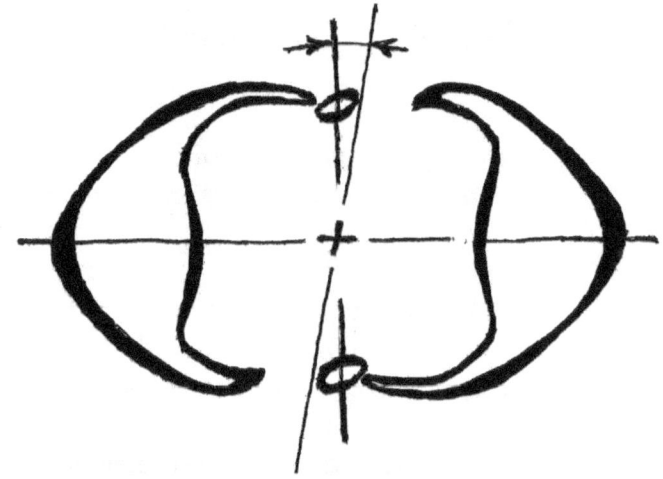

9.4 — Debris islands at polar openings float free of spinning centerline, and recoil between mold's edges and globe's centerline.

Earth-wise: Cyclically, the Earth's shell, due to its innate flexibility, undulates in tune with the rhythmic precession of the axis of rotation. Of course, there are the shell's internal strains and slippages as well as some elasticity.

As the circular polar oculi tend to fall parallel to the equatorial plane, loose terrestrial particles are checked into balanced suspension within the space of the polar oculi. These particles become centered about a newly assumed auxiliary axis of rotation perpendicular to the equatorial plane in true gyroscopic principle, that is, askew to the inclined axis of rotation of the rest of the terrestrial mass; and constantly they wobble up an down, relative to the lips of the polar openings.

Theory: By accumulation, the size of the islands of debris increases with the increase of the wobbling turbulence and begins crowding the side of the navel opening, and so the islands begin to transgress the spindle line. The tendency is to form into a new satellite globule at each navel opening, with the new globules bouncing off the edges of the navel openings with equally spaced and timed contacts tapping and bringing the whirl to a standstill.

The islands begin rolling erratically upon the edges of the navel openings (as the points of contact now are misaligned gear-like). That is, because of the physical contact, the whirling of the islets now is in the opposite direction of the general spheroid whirl; and as well these revert to spinning about the original spindle line of the globular mass.

These islands imminently come short of fusing with the navel edges and short of departing the polar vicinity. (See figure 9.5.)

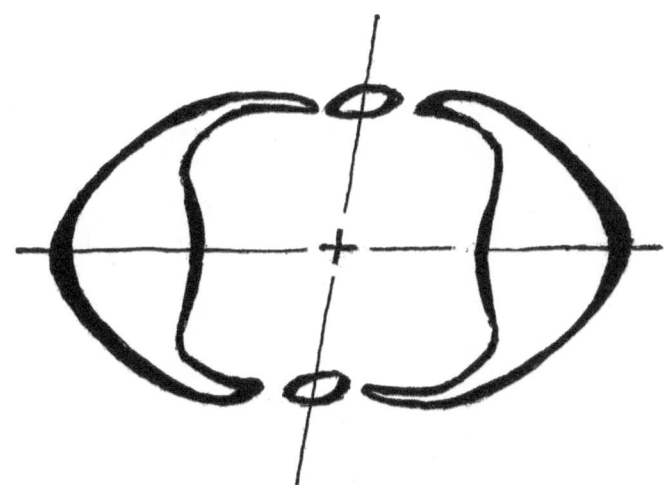

9.5 — Overcrowded islets of stray matter at polar openings, twirling oppositely, find stability again on the slanted centerline.

Earth-wise: The size of the unstable terrestrial capping islets at each of the polar oculi, along with the acceleration of the wobbling turbulence, becomes compounded in size. The appearance is as though new satellites, one at each pole, are about to be given birth and launched into the freedom of space. But the floating islands are preserved and confined between the terrestrial sub-polar mass and the locus of the conic points of the axis of rotation, as they begin rolling erratically along the polar edges, which means rolling now in the reverse direction, adapting to and maintaining a traveling tangency with the edges of the oculi.

With the reversion to a smooth vertical spin, these floating polar islands may dissipate incrementally back laterally into the mass, as they are losing their circularity.

A lateral displacement of the axis of rotation tends to dilute the floating islets, forcing their matter toward the sides of the spinning mold (mindful of the Chinese potter slapping his creation for a slight irregularity of noticeable vividness). These islets may fuse into the rest of the mass at the edges of the oculi. Or these islets or particles may escape, flying out of the Earth's atmospheric mold, tumble over and begin rotating in the same general direction the planets do, as to become satellites like the Moon, or dissipate into celestial dust.

– —— –

10 — MOLD'S VERTICAL DISPLACEMENT

Theory: An elastic globular mold is whirling about its tilted spindle line, when it is displaced suddenly downward on a parallel with its line of spindle, askew to the gyroscopic plane, but perpendicular to the orbital plane. The heavier part of the distorted mass along the circumference for a moment, due its greater moment of inertia, hesitates. The momentary distortion temporarily takes a trapezoidal shape (like a distorted persimmon with nose down). Following the thrust, it tends to compromise with its original form. (See figure 10.1, thrust's direction per slanted arrow.)

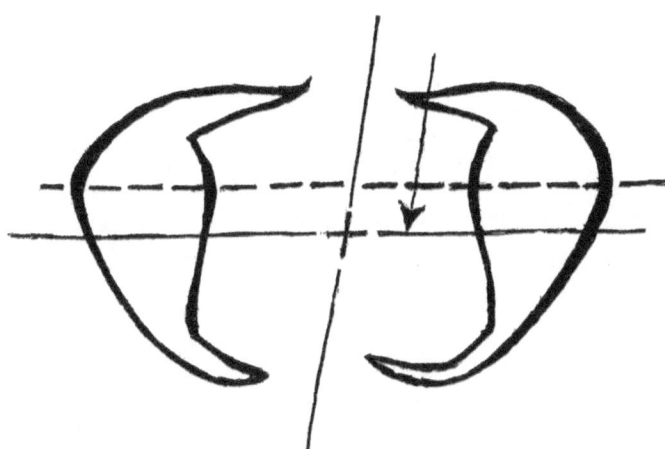

10.1 — Elongated distortion of mold with contents shifts differentially downward and sideways along its centerline, askew to orbital plane.

Earth-wise: The Earth rotating about its axis, which is inclined relative to the orbital plane, receives a southerly thrust (from whatever disturbance) parallel to its axis of rotation and askew to its orbital plane, momentarily slowing its rotation. The more massive parts of the terrestrial matter are concentrated toward the equatorial regions, due to the greater moment of inertia and the lag in their motion downward, parallel to the axis of rotation and askew to the orbital plane. This causes a sudden stressful diagonal distortion, with the lighter sections of the mass diagonally leading in the motion. The repercussion allows for partial restoration to the configuration of the spheroid. A northerly diagonal thrust acts similarly.

— ———— —

Theory: A globular elastic mold whirling on its tilted spindle line suddenly is displaced vertically downward in a direction askew to the tilted line of spindle, but perpendicular to the gyroscopic plane. Momentarily the thrust slows the rotation and creates a set of rectilinear vectors for the distribution of the force. The more massive regions resist more strongly both laterally and vertically but the less massive yield to greater irregular distortions.

Since the greater part of the mass concentrates in the upper half of the spheroid momentarily, and in the lower half there is less moment of inertia, the interior void is offset askew gyroscopically, momentarily so, causing a more noticeable wobble of disturbance in the rotation. The angularity of the section through the spheroid is greater than in figure 10.1, as the thrust is subjected as well to an irregular resistance with respect to the circumferential plane. (See figure 10.2, thrust per vertical arrow.)

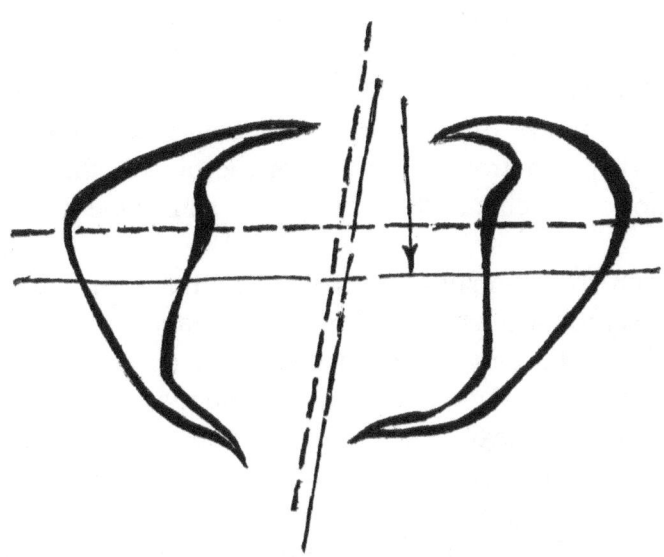

10.2 — Cyclical distortion of spheric mold with its contents shifts downward askew to centerline of spinning, perpendicular to circumference.

Earth-wise: The Earth is rotating about its axis on its precessional angle, relative to the Earth's orbital plane. The Earth receives a thrust vertically downward, perpendicular to its orbital plane and askew to the axis of rotation. Most of the terrestrial mass at the regions closest to the equator and to its axis of rotation, divide the thrust into vertical and horizontal vectors. (The resistance is somewhat greater than in the case of a thrust in the precessional direction, as per figure 10.1 above.) The vertical thrust and the momentary slowing of rotation generates a more noticeable wobble. Consequently, the Earth is subjected to greater deformities through its more complex rebound. In the process, the heavier and more massive matter beneath the equatorial zones, due to its moment of inertia, hesitatingly resists the downward thrust, while the regions around the polar oculi react quickly with the thrust. The mechanics of buoyancy in space are suggested.

These scenes happen in nature all the time. Big fish catch a school of little fish underwater by opening their mouths suddenly; as water rushes in to fill the sudden vacuum in their mouths, delivering in the little fish. Consider the globular mechanics of the big fish's mouth. As well, birds, flying reptiles, crustaceans, and all sea life are capable of educating us on the dynamics of soaring through space and water. The lips of planetary oculi, even when moving ever so slowly, perhaps control the soaring of the planets through known and unknown currents in the atmosphere, and in the interplanetary spaces.

In the cross-section of an airplane wing the upper surface bulges up and closer to the leading edge of the wing and the bottom surface is more flat. The longer path on the upper surface from front to rear of wing, facing the wind creates a lifting vacuum, which works better through the heavy air of lower altitudes. A fast jet airplane experiences greater air resistance in front of the bulging side and requires less bulging on top but greater physical force to rise off the ground and at higher altitudes it soars more easily across the skies with less effort. Dolphins are the fastest swimming of sea life; they cause the least turbulence in their wake, that is, the least frictional drag, as they re-contour their body uniquely according to their speed. The design of space ships, ships of the sea, and underwater craft may likewise be designed with differentiating adjustable flexile bodies.

— ———— —

Theory: The elastic globular mold, whirling on its tilted spindle line, suddenly is displaced downward, perpendicular, or askew to its gyroscopic plane, whether vertically or axially. Simultaneously, it decreases its speed of rotation for a moment and is subjected to a shock of greater torsion, or an exaggerated twisting wobble, where the lower half of lighter weight matter leads in rotation and the upper half drags, because of the sudden difference in the moment of inertia between upper and lower spheroids. From the compressive shock, the upper half of the spheroid mold drags in its whirling for a moment, while the distended lower half whirls faster for a moment as it dissipates the shock more readily. The torsional motions between the upper and lower halves are not the same but are compounded, tending to court a series of diagonal ruptures.

The surface of the globular mold shows the elasticity of the diagonal shear resisting the rebound from the downward thrust combined with the lesser lateral thrust, while the torsional stresses widen the circumference of the globular mold, pull the two spindle points closer together, and decrease the rate of whirling. It is like twisting a rubber balloon. (See figure 10.3, view.)

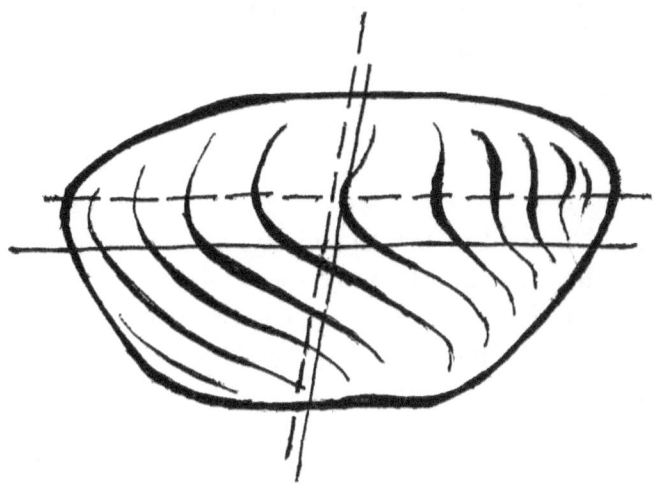

10.3 — View is of elastic shear lines, with twisting wobbles of varying moments of inertia being irregularly shifted downward and distorted.

Earth-wise: When the Earth is suddenly thrust downward, whether axially, but askew to the orbital plane, but perpendicular to the equatorial plane, the more massive sections of the Earth's equatorial zone offer a momentary state of resistance. The lighter sections, toward the polar oculi, move more readily with the jolt. The Earth's spheroid shape becomes more oblated as the widening polar regions follow through with the thrust, more readily pulling the distance between the polar regions closer together along their axis of rotation. It is like squashing a spinning rubber ball.

The new vertical irregularity, combining with the lateral irregularity and the revolving of the Earth askew along its orbital plane, yields a stronger and more complicated wobbling effect, twisting the upper and lower hemispheres against each other.

Impending diagonal fault lines for a potential rupturing become evident. A few "unexplainable" or accidental rhythmic cycles are yielded forth, cushioning each paramount wobble wave. Each cycle thereof measures today to about one and one-quarter years; it is known as the Chandler wobble.

Another factor, a general shrinkage cycle occurs in the Earth's Northern Hemisphere during "winter," while, oppositely, in the Southern Hemisphere, there is general expansion, and vice versa for the next half year, causing an irregular shifting laterally of the north and south alignments with each cycle.

– —— –

Theory: A small differentiated increase in the whirling speed of the globular mold, following the rebound of the downward displacement, is due, on the one hand, to the instant tightening of the circumference and, on the other hand, to the irregular dissipation of the momentum, distorting the mold closer to a lemon-shaped globe. The combined thrusts, the vertical and its diffusion of vectors of forces, tend momentarily to increase the size of the opening at the lower leading spindle opening and to decrease the opening of the upper trailing spindle opening. A momentary creation of vacuum sucks all into the thrust's direction downward. A capacitance for rebounding toward its original form is generated (see figure 10.4).

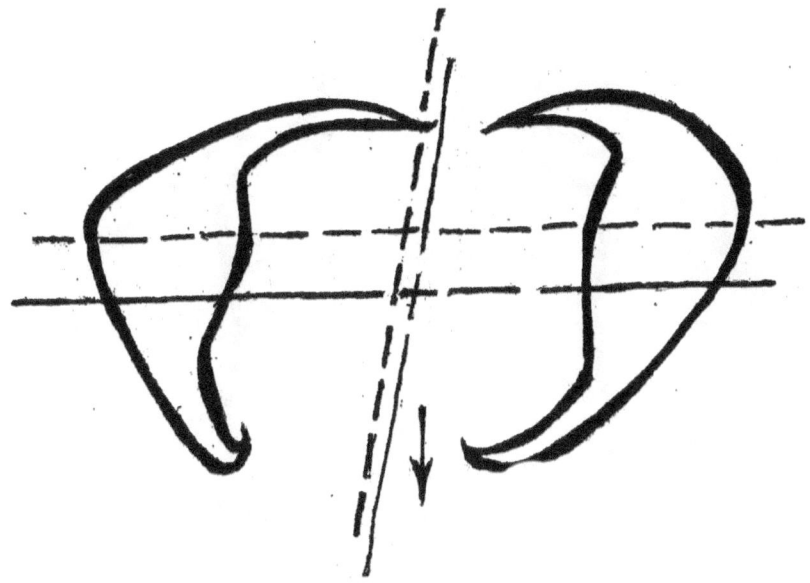

10.4 — A downward shift on a spheroid instantaneously shrinks the upper opening on centerline and widens the lower axial opening. At lowest point, right side of sketch, there occurs an accidental notch-out in the contour at bottom: please fill if convenient for you, otherwise forget it.

Earth-wise: The speed of rotation of the Earth slightly increases on the normalizing of the rebound from the southward thrust, toward becoming a prolate spheroid. The compounded motion in the stressful angular rotation of the Earth relative to the orbital plane with the thrusts combines the vertical or near vertical with the lateral diffusion of forces. Reactively, the Northern Polar oculus distends and the Southern Polar oculus expands, and having accrued capacitance of inertia, both rebound toward their original state.

— ⸺ —

Theory: The thrusts arrested, the momentum of the circumferential mass brings the centroid of the mass into normalization. The spindle openings normalize, that is, the upper now expands and the lower shrinks. Stray matter is released into the mass. Some permanent distortions are to survive as telltales. (See figure 10.5.)

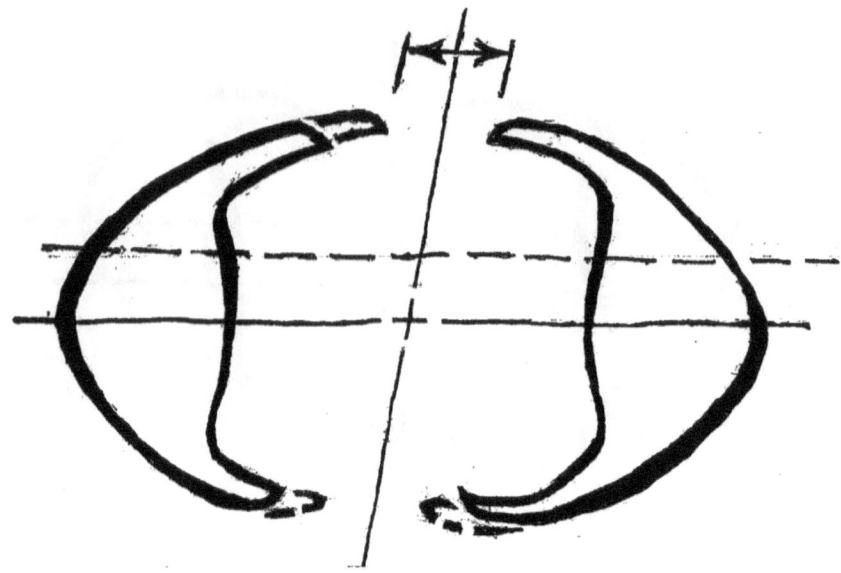

***10.5 — Following a sudden downward shift, body shape
tends to be normalized and size of two oculi tends to be
equalized again.***

Earth-wise: The islets of loose matter in the polar oculi are
dislodged and perhaps ejected into space as a missile or dust, or
swallowed into the mass of the ocular rim. Out of the loosely contained
islets bouncing around, there generates some friction transitionally,
and heat builds up around the two oculi, perhaps partially melting
icebergs, if accrued. The oculus of the North Pole expands, while
the oculus of the South Pole distends, both approaching their former
equality. The Earth regains its sphericity, but short of perfection, thus
increasing the variety of life-sustaining conditions.

— ——— —

Theory: The redistribution of the mass in an elastic globular
mold tends to be normalized internally, while the size of the spindle
openings returns to near normal. However, the slightly oblated
spheroid configuration retains the telltale due to the uneven re-
compaction of the mass, with the lower hemisphere residually

remaining slightly bloated, after having accrued filler matter during its moments of greatest disturbance. (See figure 10.6.)

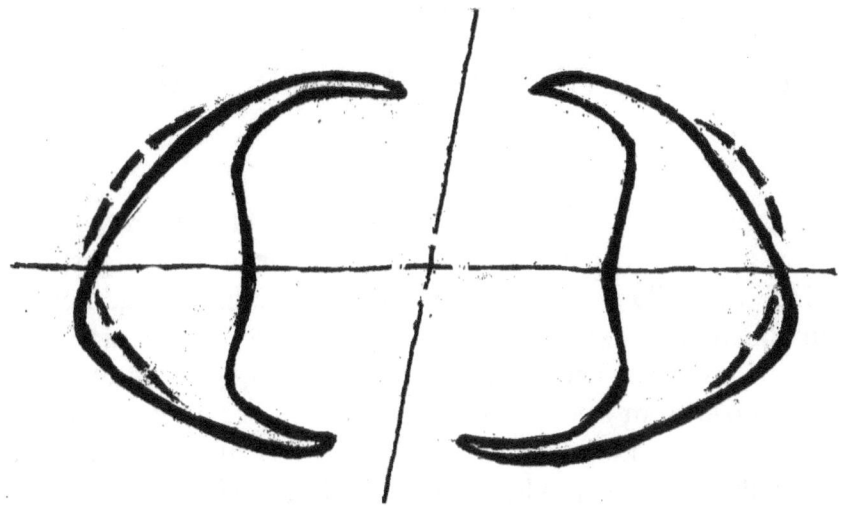

10.6 — Residual bloating of lower hemisphere is effected after recovering from a downward shift of the sphere.

Earth-wise: The reconfiguration of the terrestrial mass tends to return to its assumed former balanced oblate form and the polar oculi tend to normalize toward equality in width. However, due to the varying compositions of the terrestrial mass, the shell of the southern hemisphere retains a slight bloating.

The vertical thrusts and displacements of the terrestrial mass in the Earth's configuration, per all of the above, may be reversed as well, as from south to north. (Re-compaction of terrestrial matter may follow with greater disturbances in the southern hemisphere than in the northern hemisphere, due to the massive inequality between the hemispheres in overpowering any rebounding tendencies.) These vertical motions relate to the orbiting Earth, as each half year the orbiting Earth's equatorial plane passes above and below the Sun while on its orbital plane.

Helicopters tilt downward, in defiance of gyroscopic orientation, into the forward direction of desired propelling.

— ——— —

11 — MOLD'S LATERAL DISPLACEMENT

Theory: A globular elastic mold set upon a table on wheels, when its displacement is sideways impacted at its line of spindle, displaces its line of spindle closer to the one side of the figure, and because of inertia, a ghost axis sympathetically continues functioning at the original position. The whirling mass, assuming some elasticity, breaks loose from the original position with some temporary and irregularity stretching while in its sideway distortion. The relocated mass with its line of spindle, upon being fixed to its new relocation, substantially regains its original form. (See figure 11.1, islets at openings omitted.)

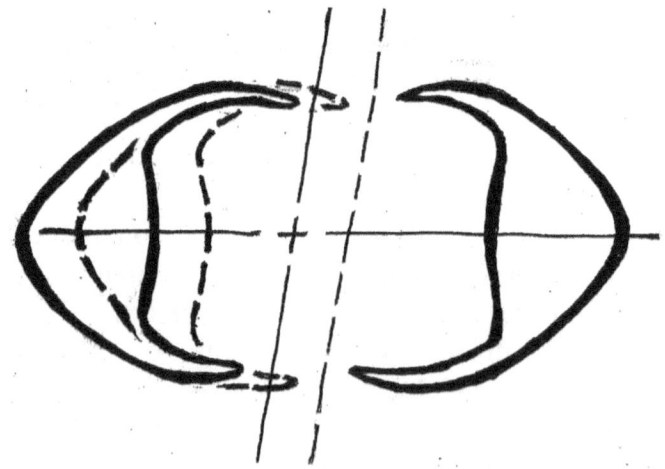

11.1 — As the centerline of spinning shifts sideways, the mass follows, and with shape restored, it leaves a shadow of the original axis.

Earth-wise: The Earth rotating about its axis is thrust at its axis laterally with the terrestrial mass, relocating the active axis of rotation along with its terrestrial shell. For a moment, in a horizontally elongated distortion, it evinces a momentary ellipse. The cause of the thrust may be celestial or terrestrial. Due to a residual moment of inertia, a sympathetic axis of rotation emerges at the original location of the axis. The two axes for a moment cause a spheroid eccentricity, until the transfer of the entire Earth is completed, and the sympathetic axis fades away. Due to a degree of stiffness of the terrestrial mass, there remains a geologic scar.

— ————— —

Theory: The spindle of the globular mold is mounted on a table with wheels to reflect the experimental shoving about. (This substantially is a repeat of the above with figure 11.1, but it is now viewed from above.)

The irregularity in the circumference occurring from a mild sideway displacement (per figure 11.1, in plan section through the circumference) yields a momentary transitional ellipse of transference and its reconstitution to circularity at the new location. The line of spindle therewith follows through, establishing the circumference's new center. The plan area within the ellipse keeps to a constant through the transformation and resettlement. An ellipse is defined as having two foci; in this case, the old center remains as the original or base focus; and the new center is the errant focus. The ellipsoidal periphery stretches and shrinks back upon resettlement. (See figure 11.2, top view, original circle with center on left, in dash line, and transitional ellipse with second or errant focus on right.)

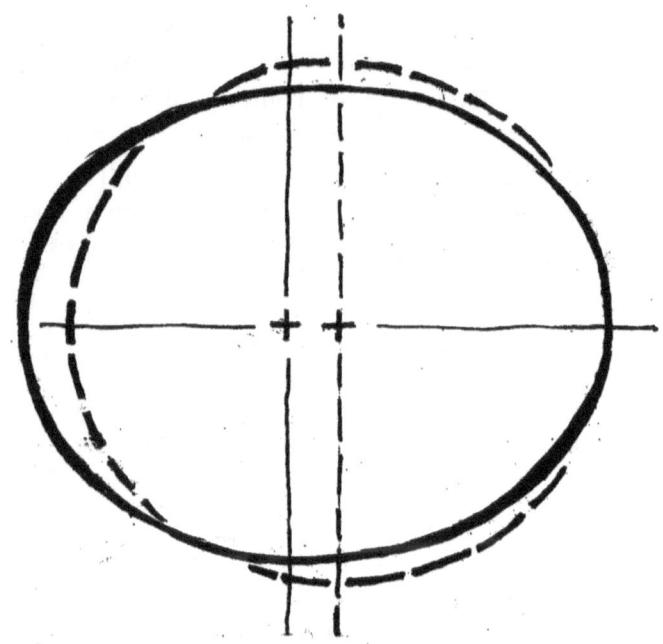

11.2 *— A rotating globule transforms to an ellipsoid globule, and upon completion of relocation it reclaims its original globular form.*

Earth-wise: A lateral thrust upon the axis of rotation momentarily distorts the terrestrial mass in the direction of the thrust. For the various moments of inertia of its various compositions and its elasticity, the terrestrial mass follows and reconstitutes itself at its relocation.

— — — —

Theory for an unusual sideway condition: Continuing with the case in 11.1, an unusual forceful displacement causes temporary irregularity of the circumference of the globule and it may experience a rebounding momentary jolt and freedom from gravity, as though to rupture in all directions the elastic mold. It appears to peel away and to discard the inside bulging ring layer as excessive. But in rebounding, it settles into a dissipated compromised position in space midway between the two foci. It reconstitutes its circularity between the two disappeared epicenters. (See figure 11.3, the top view. The irregular sideways distortion is in a dash line.)

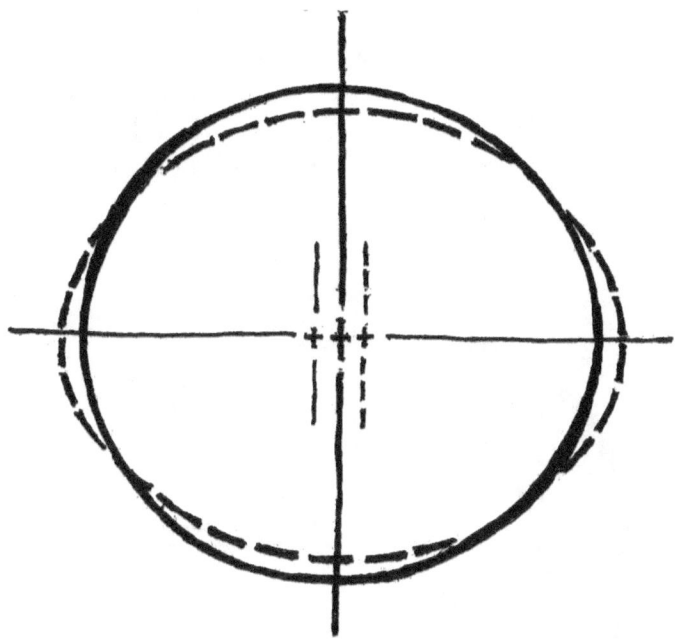

11.3 — On the left, the original position; on the right, the extreme position; and in the center is the newly settled spindle axis.

Earth-wise: A lateral thrust temporarily diverts the Earth's path of travel laterally, causing an internal distortion, which rebounds to a median position between the position of the original center and the extreme position. Upon completion of the relocation, the images of the center bearing the base location of the axis of rotation and the extreme errant position fade away. The Earth assumes its activity centered at the new compromised median position.

— ——— —

Theory: A globular mold, while rotating about its line of spindle, is thrust sideways at its spindle line and again at an angle, creating an isosceles triangle of three foci, describing an ovate in plan (egg shape, not an ellipse). The sideways movement may be pivoted off the base focus, in any direction. Depending on the intensity, duration, and consistency of the thrust, the three foci of the ovate can be thrown beyond the original domain of the circle, and the dislocation of the three spindle points may be asymmetric. But such may be accomplished through a series of thrusts and rebounds to retain integrity. (See figure 11.4, top view, dash line is the original sphere, solid line the assumed relocated elliptic distortion.)

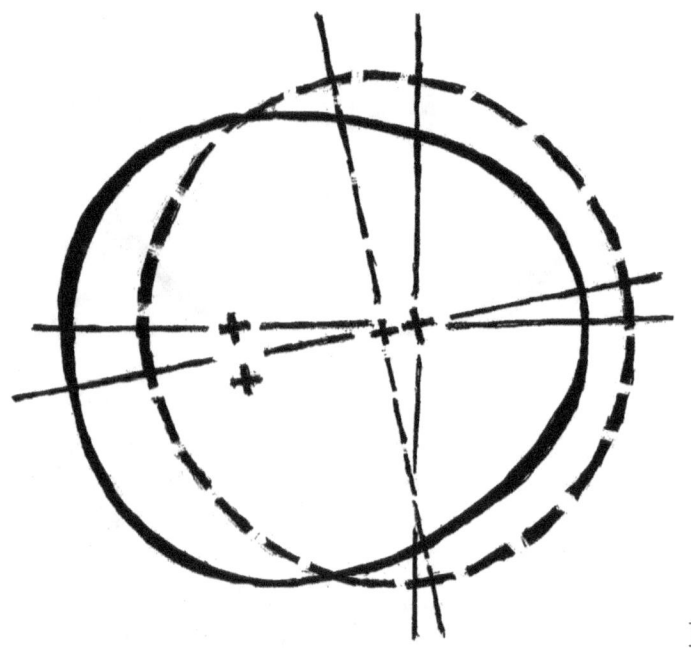

11.4 — A sphere irregularly shifts leftward, and deforms into an ovate with its first, second and third locations of its navel point.

Earth-wise: Evidence of such motions affecting the Earth shows in the misbalance of the Pacific Ocean and the rest of the continents and oceans. A view of an off circular or more complex configuration of the equator viewed from above the polar regions has yet to be made an investigative issue. However, the Moon in horizontal section is such an ovate, with asymmetry between left and right.

The three foci of an ovate need not form an isosceles triangle, but may form any irregular triangle, an irregular ovate.

When an elliptic or ovate shape is subjected to rotation, it may rotate around only one of its two foci, or while in rotational cycle, shifting to the next or errant focus and then to the third focus. Under natural circumstances it may not shift around an assumed geometric centroid of all the foci, unless for some total amalgamation, as described in the dissipating and disappearing islands at the polar oculi.

As in a musical chord of two notes, a third note is generated through the differences of the wavelengths from the initial two notes. Then on to a fifth, eighth, thirteenth, etcetera, in a series of reverberations being generated. It may be assumed with the generation of images of relocated axes in the heavenly bodies and their splitting, no three foci will fall on the same line. But these are harmonic matters and beyond the present scope.

— ——— —

Theory: The locus of parallel errant positions of a sequentially transposing line of spindle, when aligned parallel around the original (base) line of spindle, depicts a cylindric path about the new centroidal line of spindle. However, askew positions are caused by the lagging of one whirling hemisphere, which changes the angle of slant of the spindle line, depicting a hyperbolic path about the new centroidal line of spindle. The locus of moving spindle lines of positions crossing each other at some common point at the centroid of the sphere would describe a conic path about the original centroidal line of spindle.

(See figures 11.7. The spheroids are omitted, showing only the configurations generated by the locus of lines of the errant axes of rotation in each.)

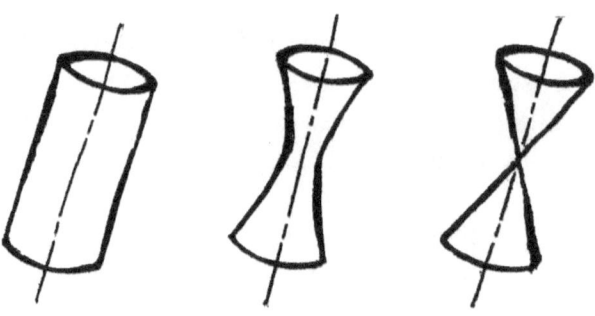

11.5 —, 11.6 —, 11.7 — Cylindric, hyperbolic, and conic shapes of locus of errant axial lines, respectively, are shown in parallel, askew, and cutting the axis

Earth-wise: As the Earth rotates about its axis, it is orbiting around the Sun as well. It is a rotation within a rotation (as whirling carousels upon a whirling merry-go-round). The Earth's axis is in precession, which is askew to the orbital plane. The precessional angle, however, vacillates over the years and ages, which means the polar projection of the axis describes a serrated (gear-like) circle in the Heavens, in the northern celestial hemisphere and the southern celestial hemisphere as well. Vacillation of the angle of precession means departing from a projected cylindric projection onto the Heavens to a conic spread (considering the center of the Earth as the focal point of projecting through the edges of the polar openings to a great circle in each celestial hemisphere.) A minor vacillation occurs about each two human generations.

The center of the northern celestial hemisphere, relative to the Earth's orientation, in our age, points to the North Star. In 2700 BC the star Draco served as the pole star.

— ----- —

Theory: When subjected to vibrations, the above three figures generate serrations in their cross sections. Figure 11.5 as a fluted cylinder, figure 11.6 as a twisted bunch of ropes, and figure 11.7 with flutes converging to naught at the conic apexes, (See figure 11.8, with fluted slopes converging to naught at the apexes of the conics with projected bases.)

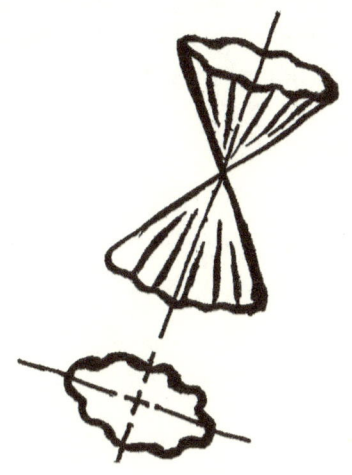

11.8 — Serrations projected as nodes at base are vacillating angles through the focus of centerlines, while the axes rotate in precession.

Earth-wise: These physical motions alone project their effect upon the constitution and timings of the Earth, her flexibility and stability, as well on interstellar and interplanetary relationships.

The path and speed of travel of the variational precession around the poles appears as a chorus of infinitesimal surges, daily, yearly, generationally, and over centuries. With each of the serrations, the circumference of the sphere is bounced in and out, while the speed of rotation of the spheric mold of the Earth appears as a series of infinitesimal surges, in compensation, and infinitesimal undulations of the circumference. Such, then, constantly stresses the terrestrial mold's structure and topography. And it raises calendric questions.

With each surge of speed and the distorting bounce on the terrestrial mold, the near-viscous mass within, while rotating, agitates, kneads, fractures, and grinds the stratified layers. In the process a viscous mass or a temporarily powdery mass is formed. The degree of comminuting is differentiated with the newly attained rigidity, or increasing viscosity of each of the layers.

The simultaneous combining of the diverse periods of oscillation of described motions, the interferences in the rotation of the sphere, and perhaps sensitively with other motions together generate into greater and lesser periods of synchronicity. The repeating combinations, including other environmental disturbances, go ad infinitum. Such phenomenon, to a noticeable magnitude, is associated with the Chandler wobble in the Earth.

— ———— —

12 — WOBBLING SPHERE'S TANTRUMS

Theory: A stiff globular mold is set fixed on its two spindles to resist any gyroscopic disorientation and any kind of distortion of the mold, with the line of spindle set upright to the horizontal plane of the mold. The contents may be liquid or like fine sand. A sufficient speed of whirling is set to keep the contents centrifugally concentrated on the inner circumferential area of the mold. As the speed is allowed to dissipate, the pliable contents begin to wave from one side to the other within the periphery, as though to force the circumference into an ellipse, while being restricted to circular containment. Both ends of the contents curled upward, climbing upward the mold, as though climbing a breakwater. And a rocking lengthening of an elliptic perimeter is thereby yielded. (See figure 12.1, ellipse rocking, while rotating, superimposed on original circle.)

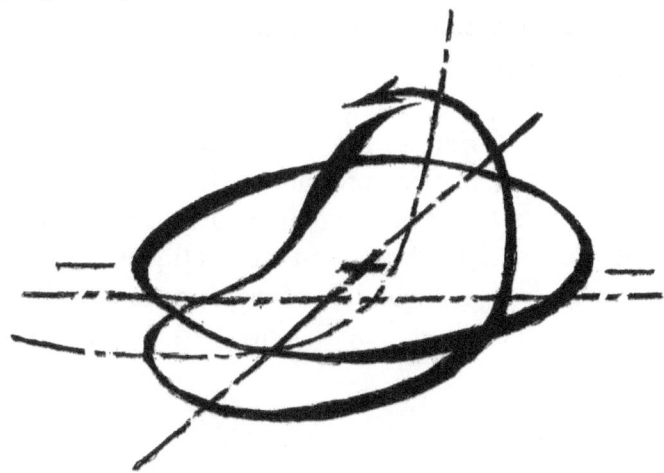

Figure 12.1 — Two sides of whirling circle with elongated ends rocking in rhythm go into a bowed ellipse restrained by the circumference.

The major axis of the back-and-forth waving of the emerging ellipse begins to turn very slowly in the opposite direction of the whirling contents. As the whirling matter continues to decelerate, the direction of the line (major axis) of waving incrementally recoils against the whirling mass. Noticeably, the waving incrementally rotates faster in the opposite direction (in back lash) of the whirling mass, until all stops and settles (see figure 12.2.).

Figure 12.2—Spiraling of warped circle is counterclockwise outward, and secondary spiral of locus of farther ends is of rotating ellipsoid.

Earth-wise: As the Earth turns, the terrestrial activity compounds and rocks laterally, and more pronouncedly so the great bodies of water. The bodies of water periodically and alternating increase and decrease their lateral spread between the continents and the islands. As the lateral spread of the bodies of water undulate, the ebbs and tides pick up destructive or quieting rhythms, usually altering the oceanic ecologies.

— ——— —

Alternate source for the theory: A supple elastic globular mold is set on a spindle with fixed but flexible contact point (universal joint) upon a horizontal plane, while the upper spindle is free to lean any which way. Sufficient speed for whirling is supplied to where the line of spindle stands upright naturally, thus causing the least resistance to the air all about and to general friction. As the speed of whirling decreases, the spindle axis begins rhythmically to lean and to rotate in precession slowly, following in the same direction of the whirling matter. The dissipation of speed causes the two opposite sides of the circle (opposite end of the bowed-up or rocking ellipse) to warp upward to absorb the dissipating momentum, pending restoration to a settled circular whirling.

The leaning line of spindle projects an image of a locus of points in a circle overhead, as on the ceiling, which incrementally enlarges with the increasing of leaning of the spindle line, due to the continuing dissipation of energy. The line of spindle increases its marginal undulation (nutations) slightly, toward the upright and to its further leaning. A cumulative capacitance of dissipating energy and recapturing of energy with each cycle of the leaning spindle introduces a rhythmic undulating eccentricity to the projected image of the circular path. The undulation alters the circumference into an ellipsoid. (See figure 12.3, with locus of shifts of orbital ellipse.)

Figure 12.3 — Elliptic wobble based on the incrementally rotating dislocated base focus impacts the globular container.

Earth-wise: The lateral distensions and contractions of the continental bodies are thrown into a slight rotation of their orientation. The rotation may be in either way depending on the direction of the major compaction.

— ——— —

More on the above theory: From the recoil of the whiplash action, the described ellipse begins to rotate around the base focus while tugging on the errant focus, in the opposite direction of the whirling mass, more commonly referred to as an "anomalistic" cycle. The reverse cycle (anomalistic cycle) increases with the decreased rotational speed, until all stops and settles.

As the incrementally rotating ellipsoid containing the rocking mass increases its eccentricity, it becomes longer as the diameter and the wall climbing permit, and the ends of the ellipsoid begin crashing, as against a breakwater, curling up and over, accelerating the turbulence. (See figure 12.4, same as in 12.3, but with more forceful rocking motion.)

Figure 12.4 — Elliptic wobble rhythmically climbs and breaks over at inside circumference of globular mold, while incrementally rotating.

Earth-wise: Though the Earth's shell is considered reasonably firm, the heat, pressures, grinding actions, and structural readjustments cause the inner matter within the thickness of the shell to react as though a rocking fluidity.

— ——— —

Practical example: In a transparent high-speed blender, observe a whirling mass of chunks of vegetables and fruits in water. Above a certain speed and a certain breakdown of the chunks of food, the whirling mass begins to rock back and forth rapidly washing up and down, even though the ingredients are substantially blended, and the twirling direction of the rocking back and forth is at a lesser velocity than the whirling mass. As the tips of the waves of the elliptical rocking waves climb the wall of the blender, they are incrementally dragged sideways, in the opposite direction, rhythmically and due to the resisting friction at the wall of the blender.

Earth-wise: The rocking waves of the oceans hit the eastern shores of the continents more aggressively because of the drag and as they climb the shores as the Earth turns. The ocean levels average forty feet higher along the eastern shores of the continents; hence, the turbulence at the Cape of Good Hope and the Straits of Magellan.

All the theoretic and earth-wise motions as presented may be recorded in the deeper geology of the Earth, awaiting discovery, verification, and perhaps improvement of the herein theory.

— ——— —

Theory: Having discussed a globular mold's vertical, diagonal and horizontal displacements, there remain the cases of three-dimensional compound displacements, but for brevity one of these ought to be discussed. It is a case of bending the active axis of rotation, as either having a universal joint, or simply by bending the axis of rotation. The axis of rotation of the upper part of the globe thus has one deviant gyroscopic orientation and the lower part of the globe has another deviant gyroscopic orientation. Or, simply the gyroscopic orientation deviates in stages from different levels along the bending axis of rotation.

The universal joint or bending peak on the axis may be at the center of the globe, or anywhere above or below the orbital plane on the axis. As the mold spins around, in each half cycle there is crushing pressure between the upper and lower halves of the globe with each half cycle, and distension between them in the other half of each cycle, and so revolving around the perimeter of the globe. This can be taken in two ways: it is a major contributor to the cause of wobbling, or it is a major factor for the absorption of all the other irregular or stray pressures, tensions and strains.

Earth-wise: All the Earth's terrestrial, watery and atmospheric activities may be unpredictable, until geology and astronomy may come to understanding the bending moment of the Earth's axis of rotation and its affects. The cause of the bending may be terrestrial or celestial, as from a shocking realignment of certain planets and-or stars.

— ——— —

ECCENTRIC
DYNAMICS

By its versatility the Earth maneuvers through endlessly changing conditions on its ordained course of travel.

13 — SPINNING TOP'S DYNAMICS

Observing the dynamics of a child's spinning top at play is quite revealing. A top is spun off at a sufficient speed, rotating in upright position in situ, let us say counterclockwise. From simple observation, as the speed of spinning wanes, the top begins leaning and traveling out to a small circular path, and it begins to enlarge its circular path, still keeping centered upon the original position when in situ. And gradually and consistently the top spirals outward as it is dissipating its momentum in surges. (This is other than the sudden axial displacements of spheres.) Slowing some more, the circular path enlarges, and it evolves into an elliptic path. The position of the axis of the top's rotation becomes split into the two foci of an elliptical orbit, the errant focus, and the original or base focus.

The orbital mechanics of motion fix upon the base focus, serving as an understood pivotal point. The slant of the axis of the spinning top, as projected upon the orbital plane, incrementally rotates counterclockwise, while traveling clockwise about the base focus, but ever slower. The slight difference in the rates of rotation opposite each other, and with their points of departure after many cycles again meeting each other, yields a "Platonic cycle" (akin to a Platonic year). As the orbiting and spinning top periodically comes closer to its base focus, accelerating and passing behind it, it whiplashes around and again wanes in velocity toward the active or errant focus, and as it passes that errant focus, it begins to wax in velocity again toward the base focus.

The waning velocity of the top upon passing the base focus causes the projected slant of the top's axis of rotation to lean over farther in anticipation of its theoretic or mechanical equivalent of its constant speed, as was before entering its elliptic path and its accrued momentum. And upon sluggishly approaching to complete the half part of the ellipse, that is, to cross the major axis dividing the ellipse lengthwise, the elliptic path absorbs the top's momentum, as the angle of the top's leaning angle is frozen in its oriented slant, while returning toward its base focus, until after the next coming whiplash.

With each elliptic cycle, and as the projected slant of the axis of spinning of the top traveling about the orbital area of the hard ground (floor or cement paving), it slowly and incrementally turns away from the major axis of the elliptic orbit – that is, the projected slant slowly turns incrementally, reorienting itself with the cyclic passing of each of the two foci. Upon passing from behind the errant focus to return, instead the direction of the momentum of the major axis is clockwise, while the projected slant of the top's axis freezes from further anticipation of a curving path and orbiting velocity, toward concluding each accelerating second half of a cycle. The errant focus incrementally is tugged and livened into turning counterclockwise for the second half of the cycle around the base focus.

In addition, the eccentricity of this *elliptic configuration* increases as the top's spinning diminishes – that is, the ellipsoidal stretches longer incrementally. And upon returning, the top's spinning eccentricity slightly pulls together, shortening again incrementally. Upon losing the last of its stabilizing momentum, for its structural incompleteness to maintain perpetual motion, its eccentricity vacillates wildly until the top falls to its side and rolls off to come to rest.

It is called *"elliptic configuration"*, rather than ellipse, because of its dynamic attempt to keep spiraling away (as in figure 12.2) and its

73

errant asymmetry with respect to its major axis, that is, the ellipse is not symmetric with respect to the major axis. The accelerating phase of the orbit commands a straightening (or flattening) of its side of the ellipse, oppositely of the decelerating phase, which appears as a truer geometric ellipse. It appears the decelerating half of the elliptic orbit is fixes on the bas focus and the errant focus is shoved laterally upon the beginning of acceleration to complete the elliptic cycle.

— ——— —

14 — PERIHELION AND APHELION

The "anomalistic cycle": As already described in chapter 13, it is of an ellipse generated by an errant focus separating from the base focus, keeping to a marginally fixed distance between the two foci; and as the dislocating energy dissipates laterally from the area of the errant focus, it tugs the errant focus of the ellipse in the opposite direction of the whirling body. This definition would explain why in chapter 11, the original location of the focus is called the original or "base focus".

The year-to-year cycle of the orbital plane incrementally is tugged and shifted counterclockwise, overshooting the zodiac standard of exactly 360 degrees of circuit, by about a second per year, while the tropical ("seasonal") cycle incrementally shifts clockwise but falls short of reaching exactly 360 degrees by a few seconds. Some of these seconds add up to a full year variously over the centuries. The anomalistic cycle is divided into the accelerating half and the decelerating half. The standard calendric measures for both rotations relate to the understood "fixed celestial equator", also known as the "celestial equinox", which, by convention in 45 BC, is set at the first degree of Ares. At that year, the celestial equinox and the solar equinox were coincident. These two rates of rotation are not equal. They relate kaleidoscopically; their bifurcating difference is a valid measure in itself of the progression of time and must not be taken as an accumulating error as per medieval "science".

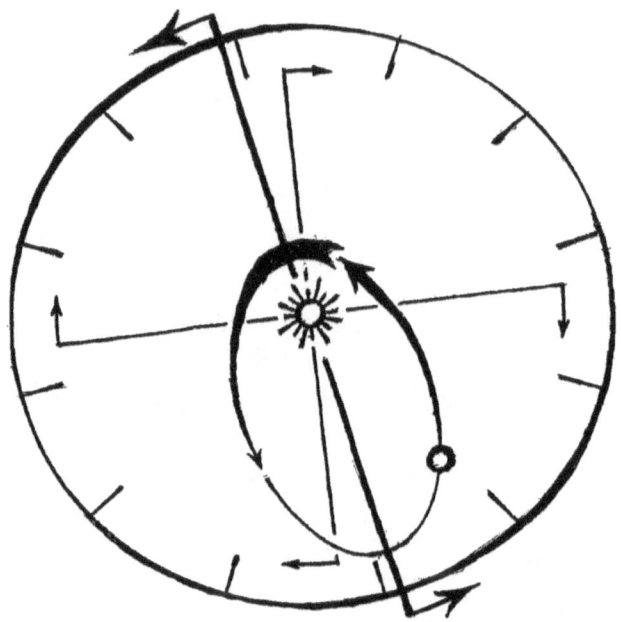

Figure 14.1 — On the zodiac field are projected the anomalistic with major axis, heavy line, and the tropical in four quarters, lighter lines.

On further theory: As the projected axis of rotation upon the orbital plane passes from solstices to equinoxes, to solstices, to equinoxes, repetitively, over the many millennia, the transitory conjoining of alignments yield a particular shock and thrust, affecting the consistency of the rate of bifurcation as well as the terrestrial conditions of the Earth's climate (schizophrenic slippage of the few factors defining "seasons") for sustaining life.

Earth-wise: The developed orbital circuit of the Earth about the Sun evolves into an ellipse, as the axis of rotation splits into a pair of parallel minor axes, one each going through each of the two foci. The errant focus over the centuries incrementally travels counterclockwise around the base focus. Theoretically, the Earth's orbital circuit is stretched out like a belt to rotate loosely around two wheels.

Perhaps the function of the base focus is as a corrective force to have the errant focus potentially conjoin into the original or base focus to be the hypothesized center, provided the Earth accelerates its rotation, contributing to transforming the elliptic path back to a circle. Then the equatorial and orbital planes would conjoin, and the axis of rotation would find its verticality again. And that means the Earth, Sun, and the star Seirios would be on the same galaxial plane again?

But can there possibly be in celestial space a perfect alignment of three or more entities? Near alignments, yes, as with the equinoxes. For, celestial space is not rectilinear, not Cartesian, but endlessly curving. (In Mythology is mentioned that Man migrated to Earth from Seirios.)

In its orbital path, the Earth crosses the major axis twice, once at each end of the ellipse. These two points are the perihelion, at the end closest to the Sun, and aphelion, at the other end farthest from the Sun. The Earth, orbiting counterclockwise, whiplashes around perihelion, decelerates toward aphelion, is slowest around aphelion, and accelerates toward the point of whiplash again, mindful of the "spinning top," completing an anomalistic cycle, or year. (See figure 14.2, reoriented from above figure.)

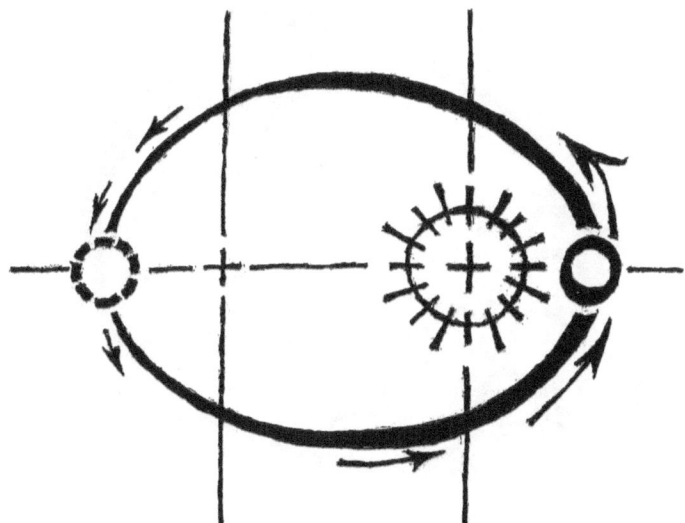

14.2 — Earth's orbit quickly passes perihelion, its velocity waning, and slowly passes aphelion, again waxing to complete the cycle.

The distance between perihelion and the Sun is 91,500,000 miles, and the distance between aphelion and the Sun is 94,500,000 miles, the average distance being 93,000,000. The distance between the two foci of the orbital ellipse is only about 3,000,000 miles. Note, the ellipse is quite slight, with an eccentricity vacillating around 0.0167, and note how dramatically our "seasons" are affected. These distances fluctuate almost a million miles up and down; therefore references no longer strive for accuracy.

The average speed of the Earth's orbital travel is 66,500 miles per hour (72.9 miles per second). Imagine a spaceship returning and trying to make a landing, and nevertheless they are accomplishing it.

From the point of aphelion, as the Earth renews its acceleration, in reaction it tugs the point of aphelion, or rather tugs the errant focus, incrementally counterclockwise. The tugging is in the opposite direction of the rotation of the equinoxes and solstices, which means the circuit around the zodiac belt counterclockwise overshoots the 360 degrees of a circle. This circuit defines the anomalistic year, which requires over 100 centuries to accomplish one complete circuit of perihelion around the zodiac belt (see figure 14.3)

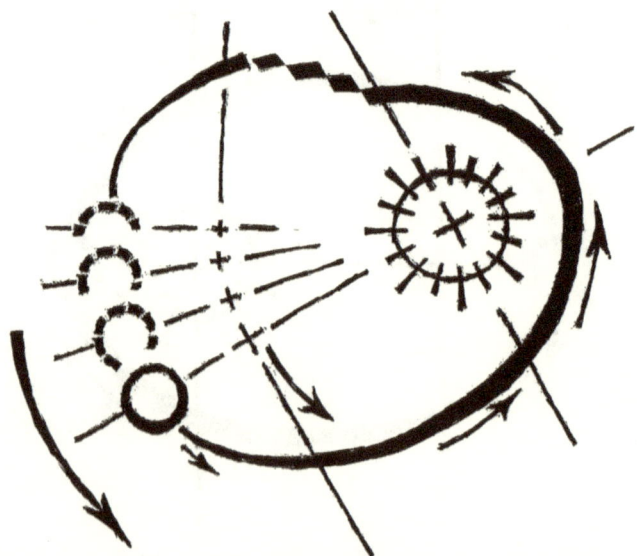

14.3 — Earth's orbital ellipse twists about its base focus clockwise, slowly in opposite direction of rotation being tugged by errant focus.

The swirls in water drains of plumbing fixtures are counterclockwise when north of the equator, and clockwise when south of the equator. Oppositely, trade winds north of the equator generally swirl clockwise, and counterclockwise south of the equator. Now, the Earth and all planets in our galaxy rotate generally counterclockwise. Ergo, could there be an immediate galaxy directly south of our galaxy, when looking from north to south, wherein swirls and trade winds respectively rotate oppositely to fill a "gear-like" celestial balance? Or, similarly, directly north of our galaxy?

An ellipse is defined as having two focal points; if the orbital ellipse of the Earth's path were perfect (as so far demonstrated it is not), there would occur no anomalistic cycles. More specifically, as the Earth passes aphelion, it hooks and jolts the errant focus laterally an infinitesimal amount, creating a hardly distinguishable ovate and reorienting the major axis. And upon accelerating after immediately passing the second errant focus, the Earth's path tends to straighten until jolted into whiplash around the base focus.

(The eccentricity of an ellipse lies between 0 and 1; of a circle, it is zero; of a parabola, one; and of a hyperbola, more than one.)

The decelerating half of the orbital path, from perihelion to aphelion, is the true half of the ellipse. And accelerating half, from the aphelion to perihelion, is the flattened ovoid half of the orbital ellipse. Notwithstanding, aphelion covers a slight spread of a pair of errant foci (see figure 14.4).

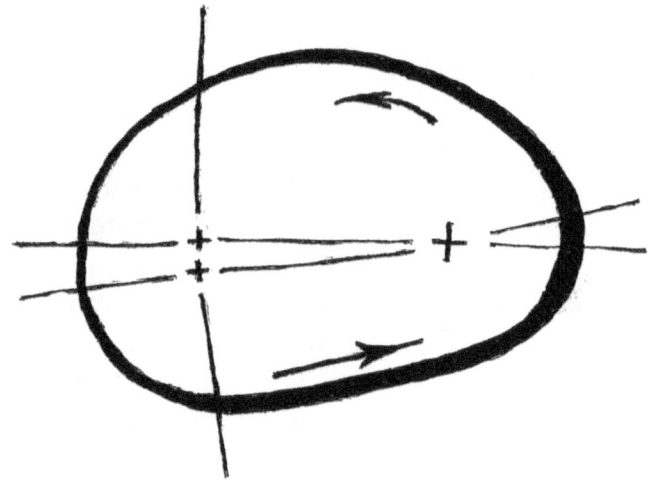

14.4 — Perihelion to aphelion decelerates in elliptic path, short arrow; aphelion to perihelion accelerates in ovoid path, long arrow.

Because of these anomalies, each quarter being unlike in shape and area (highly exaggerated in 14.4), the lengths of the four "seasons", as defined by the solstices and equinoxes, are in unequal quarters.

It would require about 111,300 years for the errant focus, or the pair of errant foci, to make one round about the zodiac. It would require 20,000 years to lose one set of "seasons", as the "seasons", are on another criterion of progression, needing yet to be defined, also make their round about the zodiac.

— ———— —

So far, an outline is suggested for the possible discovery of perpetual motion, propulsion and impulsion, to be operatively affected, and perhaps with a very minimum consumption of fuel.

The ovoid cross section through the Earth's orbital plane, with the three foci undulating relative to each other, rhythmically stimulating a whirling, may be contributing to a pumping action, by which the planet's rotation approaches perpetual motion. (It is mindful of a kind of bicycle with a rear wheel having an offset axle instead of foot pedals, where the rider gently bounces his weight up and down for power as he rides along, as well as he rhythmically controls his speed.)

Let the toy top, or the Earth, be a spaceship that breaks from its trifocal rotating ovoid and either falls to the floor to rest or sails off and away freely to another region of the galaxy, or to another galaxy altogether. But retaining sufficient momentum, it could affix itself, or be affixed, to another solar-like focus. That is, if Man comes from another astral system, this would open the probability that he can go to another astral system. What could "removed from paradise" imaginatively mean? Was it possibly to or from another planet?

— —— —

15 — ANOMALISTIC SWELL CYCLE

As the orbiting Earth approaches perihelion, the pull toward the Sun is for the moment greatest at the Earth's equator because of the nearest part to the Sun. The equinoxes would particularly be altogether stronger during those eras when they would fall at perihelion and aphelion, rather than on the solstices. The Earth remolds itself toward greater oblation, with its diameter increasing and its rate of rotation about its axis decreasing. Even the Earth's circumference is minutely and briefly knocked from its circularity into ellipticity upon the moment of whiplash past perihelion; as suddenly it relieves itself of the extreme of the Sun's pull, lasting for a day or less in each year.

In AD 1200, the projected axis of rotation upon the orbital plane and the line of the solstices were aligned, collinear with the major axis of the ellipse. North Europe experienced its warmest winters in the AD 1200s. And 11,000 years ago, as documented in the Orphic poems, the major axis of the anomalistic ellipse was aligned with the equinoxes, which means the one equinox was the coldest day of the year and the other the hottest; and in both cases in spite of the length of day being the same. The Mythology on Phaethon and his flying chariot alludes to this.

The Earth's rotation is slowest while crossing past aphelion, delayed somewhat due to weaker momentum, as the Earth is returning from its cyclical oblated shape. The slower rotation allows absorption of more heat from the Sun, even though more distant from the Sun. (Rays convert to heat upon contact with any surface.) And at the same time the heat compounds within the Earth's shell in the buildup of heat due to the momentary increase of centrifugal and centripetal forces. The difference between the Earth's centripetal and centrifugal forces, as it stands to reason, is strongest during perihelion and weakest ay aphelion, especially when hosting the equinoxes.

Around perihelion, the twenty-four-hour day is slightly longer (that, too, is elastic) because the Earth's rotation by celestial friction is restrained during the whiplash. Therefore, February has only twenty-eight or twenty-nine days, although fitting in the thirty degrees of arc, but with less travel time along a shorter arc length than the average arc lengths of the ellipse. (Although twenty-four-hour periods may be longer and shorter and not exactly fitting into thirty degrees of arc, the average throughout the year defines the twenty-four hour mechanical time clock.)

Toward and around aphelion, the reverse is set into motion. At aphelion, the day covers more arc length and more time in each thirty degrees around the zodiacal belt in its orbit. Hence, July and August consecutively have thirty-one days. And it is not because of Emperor Augustus's jealousy for Emperor Julius Caesar having one day more in his month, as we are taught in secondary education.

It is like one swinging a sling, while rhythmically tugging it to increase its speed and centrifugal force to sustain its momentum. These anomalistic yearly swelling cycles, regular but subtle, serve as a contributory pumping action incrementally to the northward continental shift at perihelion.

Of course, there is heat at perihelion from the Sun too. While passing aphelion, the differential of temperatures in the Earth tends toward being evened out; but while passing perihelion, the differential of temperatures tends more toward extremes: where hot, hotter and where cold, colder. The manifestation of this in our era is in the summers and winters in the Northern Hemisphere temperature-wise experiencing greater extremes. In the Southern Hemisphere, instead, the winters —which fall at the time of the northern summers— are longer by about four days, but not as cold.

The phenomenon lies, on the one hand, in the law of rays losing power by the square of the distance. On the other hand, proportionally the dimensions of the Earth in volume become somewhat bloated at aphelion and hence are greater, which is when the distance of the Earth from the Sun is at its greatest.

Therefore, the Northern Hemisphere, in our era, is subjected to greater internal motions of expansion, contraction, and shifting, which are contributory to the pear shaping of the Earth, whereas materials tending to rest accumulate in the Southern Hemisphere, but short of Antarctica. The measuring is in the time to traverse from each of the solstices to each of the equinoxes, these quarters being of unequal time and lengths, although each is of a ninety degrees part of the zodiac (see figure 15.1).

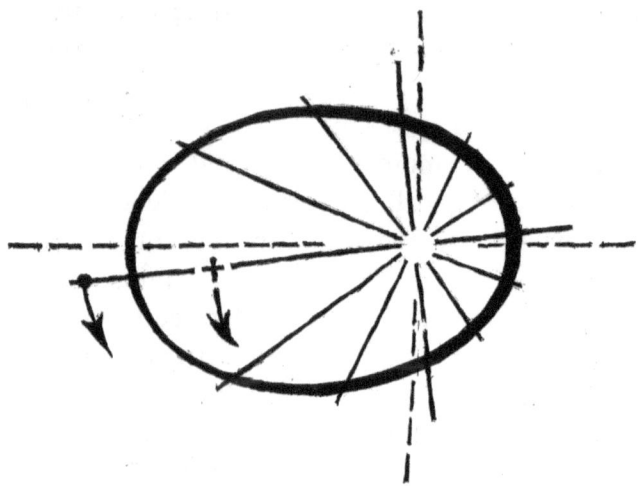

15.1 — Orbital ellipse in twelve segments: each is thirty degrees but of differentiated arc lengths, segmental areas, time lengths, and orbit speeds, in each twelfth of a year; orbital cycle is counterclockwise.

In our era, the bouncing action of the Earth, relative to its orbital plane, undulating below and above, and the Earth's turning about the barycenter with the Moon, supplies the incremental jolts daily, or twice daily, contributing to the general northward continental shift and other impending motions.

— ——— —

16 — UNIVERSAL ECCENTRICITY

More or less, no celestial and terrestrial motions are straight or circular, they are curved, elliptic, warped, refracted, and so on in varying degrees and in nonalignment. All is in undulation and always in a state of flow. The Earth in plan section through its circumference is slightly eccentric, and the Moon more so, both ovate, rather than perfectly spheric. Lines and planes are not straight but bent or warped. Galaxies are elliptic and twirling. Lines are naturally aberrated as they are directed through "space". The lines and ellipsoids are askew to each other. Nuclear activity is ellipsoidal, etcetera.

Supposed that all these lines through space are alive with sundry energies, then of what near "solid" web of convertible energies is our universe composed? Could we one day walk all through the universe on these lines, like loggers do breaking up log jambs in rivers without falling and getting wet?

Perfect shapes would mean death. Irregularity and non-perfection means living activity, perpetually pumping, agitating, growing, speeding, reversing and so forth.

Each of the activities, universal, galaxial, and terrestrial, relates to all the undulating motions, and each of the many particular activities and motions seriously affect more so certain flora and fauna. Man's activities, vis-a-vis elements and energies, bring Man to the problem for establishing a common calendar, a problem because all criteria become confusingly multifaceted. A different calendar, rather a different ephemeris, for each order of the way of life for mankind, including his cultural activities, is of necessity established, even under disagreement and protest. Rather, according to his cultural necessity, one of the four "ephemerides" usually becomes his standard calendar.

Almanacs are not the same for each year and not the same for all that the Earth bears and brings into waxing.

"When Man fell, the entire universe fell!" Even the primordial cycle of time "fell." If so, then what can a "thousand-year peace" mean?

— —— —

Is it possible for the axis of rotation of a planet to lie upon its orbital plane, with an equatorial plane perpendicular to its orbital plane? For clues and causes Mythology ought to be more carefully scrutinized.

The axis of rotation of the planet Mars is tilted past ninety degrees, with its north pole crossing seven more degrees south of its orbital plane. Because it does not lie upon it orbital plane, it may be deduced that the axis of rotation somewhat spins while orbiting around the Sun. With each cycle of spinning, sunlight enters the one polar opening, ricochets through the hollow of Mars and escapes through the other polar opening. The axis of rotation of Mercury stands upright.

— —— —

EARTH'S
STRUCTURE

Earth is providentially designed to keep restructuring itself in conformance to an ever-changing environment.

17 — EARTH'S WHIRLING MASS

Let the elastic spheric mold be the Earth's surface with its immediate atmosphere and its various electromagnetic charges, let the pliable mass in the mold be the sub-surface mass of the Earth, and let the inside ellipsoid, or barrel-like, or any kind of internal configuration, be the great hollow of the Earth. Let the poles be capped with ice and-or other matter caught in the wobble at the polar oculi, and kept suspended in place. (See figure 17.1.)

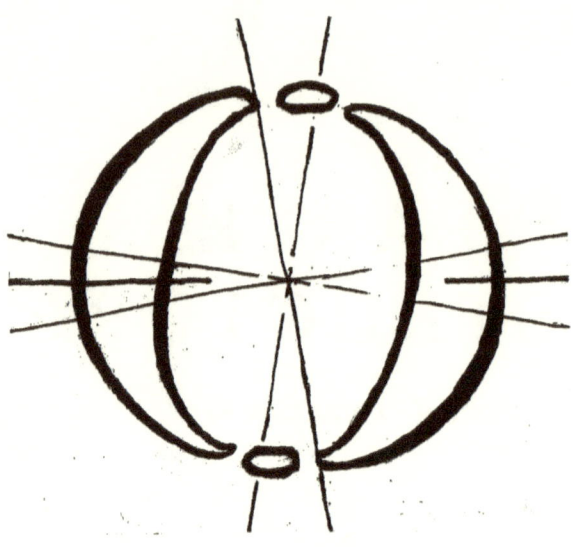

17.1 — Earth is shown hollow with loose or floating polar caps.

Human psychologic need would want the hollow of the Earth to be small and tunnel-like, and the mass of the earthly shell to take up a substantial amount of the volume of the ellipsoidal configuration. Conceivably, however, the Earth's shell can be quite thin, leaving a vast internal vacuum. The thinner it is, the more easily can be explained the kite-like properties of the Earth, as well of other planets and stars in heavenly suspension, and many of the internal terrestrial phenomena.

It is not simply a matter of only celestial winds to float the Earth in outer space, but perhaps of other yet-to-be discovered forces and energies as well. Even should the Earth approach an eggshell thickness, the inner and outer atmospheric forces, describing the molding energy for the Earth, ensure its shape, stability, molding, and flexibility with its vacillating and elastic but tough configuration.

The understood mold that holds the Earth from spinning out of shape is for the most part nonmaterial, a form of energy, which may include among the internal forces the centripetal, centrifugal, gravity, magnetism, eddy currents, various atmospheric and other known and unknown forces, as well as sundry external forces acting upon and "hammering" the Earth into shape, from within and without. Such forces not only preserve the integrity of the Earth's shell, but they also buffer the Earth against other possibly invasive planets, comets, etcetera. As well there are the forces, which induce the Earth to conform in maintaining certain characteristics, so that she would fit well in her allotted celestial setting. The Earth may reject or accept matter and energies from beyond, and eject matter which would be better suited for the outer universe, than burdening the Earth.

Perhaps, opposite to our common deliberation, a more massive shell, but with an appropriate configuration, would encompass capacitances tantamount to having anti-gravity properties interacting with outer space — speculatively, a more massive and lively shell paradoxically may lighten the Earth's weight.

Planetary collisions are a near impossibility. But comets and celestial debris are another matter, perhaps of less danger than imagined. Many of these substantially burn out or dissolve upon penetrating the Earth's atmosphere.

The degree of distortion from sphere to tangerine to apple-shape is related to the constantly differentiating centripetal and centrifugal forces, the mass of varying constency and molding, the involved varying rotational speeds, the unbalanced accumulations on part of the Earth, and the disappearing of the "Pacific Continent". Several other exterior and interior forms of energy prevent any implosion or explosion of the Earth's shell. The centripetal forces, and perhaps the voluminous atmosphere, although light in weight, press inward on the outer surface of the Earthly shell. The Earth is kept in perpetual spinning by nature's sundry forces interchanging, complementing, thrusting, and pumping; these include celestial, solar, atmospheric and internal, and the greater celestial government. And are there other causes yet to be discovered? The nature of such government is, as is "all that is", a creation of Divine Providence, preexistent to all other matter and their alternates in kinds of energies.

Note, when discussing power, there are the created energies and the Divine Energies, which are not created. The latter, though effective upon all that is created, is another subject.

— —— —

18 — EARTH'S MANY LAYERS

Commonly, and perhaps a psychologic necessity, the term shell or crust of the Earth is imaginatively thought of as an ever-crumbling and shifting cover over a more solid base or core. Also, it is thought that the oceans fill the upper voids, completing the covering of a lower tectonic shell, as if the waters are the lowest and heaviest part of the atmosphere rather than an integral surface of the part of the planet Earth, notwithstanding that some waters evaporate into the Earth's atmosphere and precipitate back upon the Earth.

"He separated the waters from the waters." Waters, for practical purposes, also must be considered as part of the mass forming the shell. Hence, and by default, the theories of continental plates shifting and folding to form the mountains and valleys arise. In conjunction, the stresses, strains, re-figurations, expansions, and contractions contribute to the contained, controlled, and compensated "instabilities" in the Earth's flexile shell.

The question arises, whether the continents float on water or on some muddy, oily, or graphite-like medium, which may be a part of an all-inclusive ocean bottom. Or, do the oceans fillers between the continents lay over a more mechanically rigid structured substratum? That, these are constantly subjected to a moving friction and fracturing is neither supportive or refutative of a hollow Earth. The oceanic mass and terrestrial shell are also subjected to the same centrifugal and centripetal forces, as are the atmospheric and exospheric layers. That the continents float fits better in the scheme of endless stratification, and it mitigates any shocks and severity of periodic "destructions", as from earthquakes, hurricanes, and cosmic bombardments.

The shell of the Earth, in other words, must be of several thin concentric shells, of discontinuous and integral plates, each of a different density and consistency. Each of the shells more or less is thickest at the equator and gradually thinning toward the polar regions. The shells need not be symmetric and in comparable patterns, as they are rhythmically being shaved down to fit, and a balance is always to be found. But such balance may be disturbed, even violently, when the Earth's given data are altered and the plates break up and fuse more violently. (See figure 18.1, imaginary.)

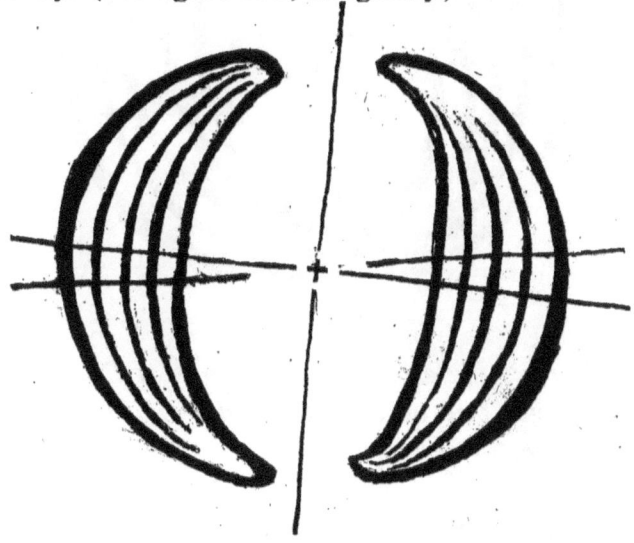

18.1 — Earth's plates in theoretic concentric balance, each is thickest at equator and thin to naught toward the polar openings.

Each of the Earth's layers, or shells, slides everywhichway and twirls about, independently of its neighboring layers, each laying on the other with variegating wear on each other. The Earth is constantly undergoing stress and distortion, bringing a misalignment of the axes and polar oculi. This means the one or the two polar oculi at times may be closed or obstructed by the endless state of varying misalignment. Because each of the layers is not a perfect sphere, neither on the outer nor the inner surface, its sliding is forceful and generates friction, distortion, and fragmentation upon each other, and a recurring unevenness to the shape and thickness throughout the earthly layers; some break, ride over, or distend and crumble. At the poles, some crumbling takes place so as not always to allow a perfect closure for any extended time. (See figure 18.2.)

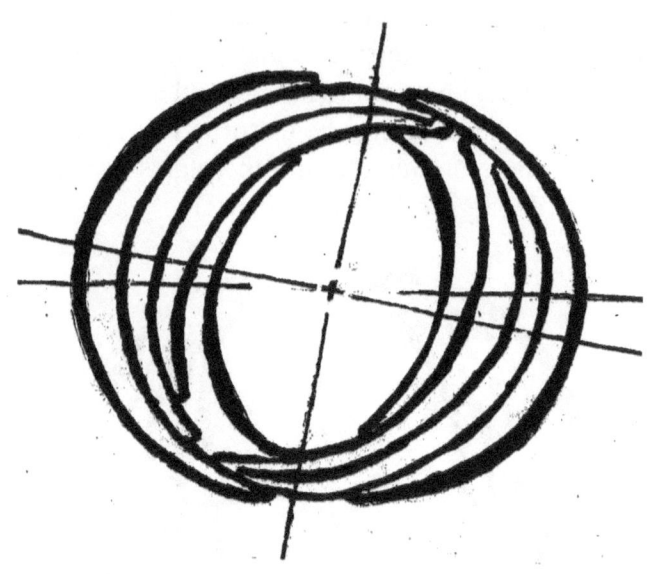

18.2 — Earth's plates at times slip out of line past each other and over one or both the polar oculi, temporarily covering them.

Each layer is composed of different natures of matter, elements and charges. Accordingly each layer is affected by the forces or influences of another planet or other heavenly body, all are undulating as their cycles relate to the Earth's movements to and fro, and to the external forces. Periodically, the plates slip to conceal the oculi.

Influencing examples: the Moon influences the Earth's oceans; Mars influences the iron layers and deposits; the Sun influences the petroleum deposits, plant oils, etcetera; and other things in the Heavens influence each of the tectonic plates and so forth. All cause the varying capacitances of energies. We live in an interesting universe, far from isolation and indolence.

It may also be ventured to say that each layer of the shell, above and below the Earth's surface, might have its own gyroscopic orientation with reference to a given particular heavenly body. Mechanically speaking, the exact point of tangency of any part of a shell to any part of an adjacent shell is hardly repetitive.

— ——— —

19 — CONTINENTAL SHIFT NORTHWARD

Each year during the Northern Hemisphere's winters in our era, all the continents receive a northward incremental thrust when passing closest to the sun, "perihelion", as the northern hemisphere leans away from the Sun, while the southern hemisphere leans into the Sun, also receiving a stronger thrust toward crossing the equator northward because of the changing curvature.

Incrementally, the continents are heading for a constriction and pileup nearer the Arctic Circle, short of the Arctic Circle, because of the Earth's curvature reversing, once passing to the north of the equator, while the continents are yet traveling away from the Antarctic Circle. The cause is in the particular whip-lashing characteristics of the Earth's orbit in our era at perihelion, and the vectors of thrust outward and upward, according to the changing position of curvature.

(See figure 19.1, large arrow shows direction of whiplash passing perihelion; small arrows show direction of upward thrust of continents, and large arrow the whiplash.)

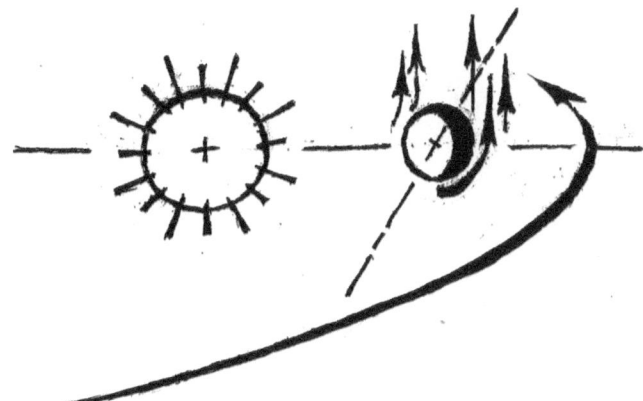

Figure 19.1 — Earth's elliptic path passing perihelion thrusts all the Earth's continents northward.

Glaciers and icebergs, however, are prone in our age to being beckoned southward, rebounding from the whiplash past perihelion. There is friction between the glaciers and the surface of the Earth, but under pressure momentary heat movement is allowed them to slide as if on water. The waters seek a counterbalance of the weight between the two hemispheres. The same forces , in our age, attract the land and icy masses during aphelion, but for the Earth's greater distance from th Sun, these forces are less effective, while the watery and icy matter are more easily drawn, especially in the moment past whiplash.

When the equinoxes, in another age, are approaching alignment with perihelion and aphelion on the major axis of the elliptic orbit the continental shifting slows, comes to a halt. And when passing beyond then the shifting reverses it course of travel to southward. Accordingly the icy matter reverses its path, oppositely, as well.

— ———— —

The projected orbital ellipse of the Earth, when projected to lying upon the Earth's equatorial plane, it trigonometrically approaches a circle.

— ———— —

20 — CHANGE IN CONTINENTAL SHIFTS

The general northward thrust of the Earth's continents drifting incrementally northward changes with each year, along with both subtle and perhaps radical differentiations of the characteristics in the Earth's orbit.

When the angle of the axis of rotation projected upon the orbital plane, in some other era of long ago, is progressively turned toward alignment with the minor axis, the reactions would be as follow:

The one solstice of the Northern Hemisphere at such time leans farthest away from the Sun as the Earth crosses the minor axis of the orbital ellipse (instead of today's near parallelity with the major axis) and closest as the Earth again crosses the minor axis in the second half of its yearly orbital cycle. These points on the minor axis in our era bear the mean speed of the Earth's traveling velocity between the points of acceleration and deceleration, most accurately so in the 1200s.

The two equinoxes, through each 5,225 years, shift around from the minor axis to fall at perihelion and aphelion in those years when the solstices align with the minor axis. The equinoctial line then in another 5,225 years shifts around to be collinear with the major axis (not the minor as in our era) of the orbital ellipse. The northward shifting of the continents is then neutralized to a complete halt. The climatic conditions of the entire Earth are seriously altered in character. Wastelands would become valuable to civilization, Man's already developed lands may become deserts, bodies of water would dry up, new bodies of water and rivers would issue forth. Old mountains might sink into the oceans or lands; new mountain ranges lying in other directions would rise from the oceans or lands. The peak of cold weather and the peak of warm weather will occur during the equinoxes (not on the solstices as in our era). In other words, at that time, the coldest day is not the shortest and the longest is not the warmest (as we nearly experience in our era) but are of moderate transformational temperatures.

The solstices with the equinoxes rotate with respect to the major axis of the orbital ellipse ninety degrees, that is, in each three-month period, each 5,225 years. The projection of the axis of rotation of the Earth upon the orbital plane turns from perihelion to aphelion every 10,450 years, where the North Pole in AD 22,450 (1200 + 10,450) will lean toward the Sun and away from the South Pole).

Therefore, what is a "season"? It must be redefined. The "tropical year", advances by three months in relations to the anomalistic year, in each 5,225 years. Nature will redefine the seasons, from coldest and warmest on the two solstices, while the two equinoxes will bear the longest and shortest days with moderate temperatures for both. The whiplash effect in passing closest to the Sun takes place on the Earth at the instant of the one equinox (rather than the solstice).

– ––––– –

As in the study of linguistics, a bifurcation between the etymology of a word and its present semantics is lexicologically noted: the definition of "season" bifurcates not for Man's lack of linguistic discipline but from the changing physical qualities of what constitutes a "season," which is not clearly extant in Man's history, and incorrectly defined in the dictionaries.

Specifically, when the Earth's hemispheres lean, respectively, closest and farthest away from the Sun during the equinoxes, then all the continents tend to gather toward the equator. And the atmospheric garbage tends to collect into a disc in the atmosphere on the equatorial plane just above the Earth's equatorial belt, mindful of Saturn's ring.

A key to the vastness of the Pacific Ocean and to the supposed existence of a great Atlantic Island may be due to a position of a wobble wave, perhaps hugging the one side of the Earth for a sufficient amount of time.

As the Earth's leaning axis of rotation will bring the Southern Hemisphere farther away from the Sun, instead of the Northern

Hemisphere, in a 180-degree turn of the major axis of its orbiting ellipsoidal circuit, requiring about 10,450 years, a precession of about six months, the continents will begin to reverse their flow to southward and travel incrementally toward the South Pole. The whiplash at perihelion, with the North Pole leaning toward the Sun, will then be thrusting the continents southward. And at aphelion a secondary whiplash will aid in throwing the continents outwardly toward the South Pole.

Should the Earth return to a more perfect sphericity, the continents would tend to be distributed evenly toward both poles.

Perhaps there is another motion, wherein the ellipse's focus upon the Sun deviates off center from the Sun. Depending on the direction of deviation and whiplash, the thrust can be either dampened or compounded.

Here are some of the natural conditions for the so-called "evolution" of the flora and fauna, but which in another 10,450 years may, or may not, evolve to their previous state because of other cyclical celestial rearrangements.

— ———— —

21 — INWARD-OUTWARD ROLLINGS

The movements of the outermost and innermost plates and fragments seem to be the more active than those sandwiched straddling the median thickness of the Earth's shell. The solid sections of the outer surface, in our era, tend to disappear into the North Pole as the Earth passes perihelion, and the inner surface of the Earthly shell incrementally tends to be drawn out to exposure through the South Pole at aphelion. (See figure 21.1, plate movements only.)

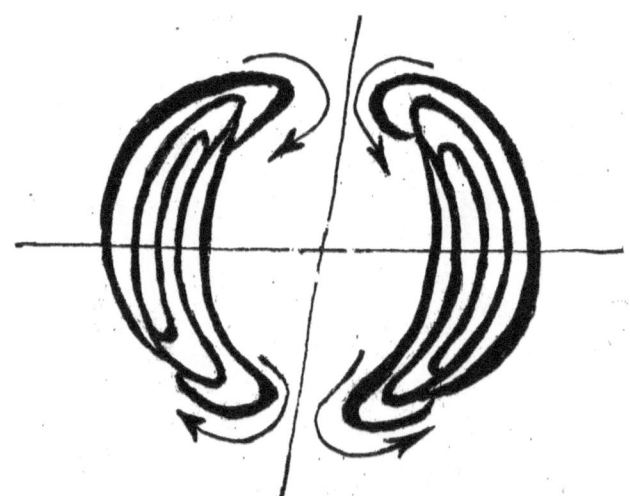

21.1 — Outer plates shift northward, breaking, crumbling, and being swallowed at North Pole. New lands emerge from the South Pole.

Each of the layers forming the crust of the Earth is composed of an aggregate of internally slowly rolling solid and softer matter, rather than a single inflexibly solidified mass, each perhaps of a varying consistency, or formula. The layers are kept separated by the unique specific gravity of each. Layers emerging to the surface of the Earth and layers disappearing into the inner surface through stresses and strains undergo readjusting cycles, with partial or entire reconstitution of matter and structure. Perhaps this is due to an uneven speed of rotation and the slippage, or rubbing against each other, between several layers of the terrestrial shell, and of the centrifugal forces being greatest at the equator, visiting the Earth with interminable distortion.

Within each of the layers of the Earth's shell, there take place dissolutions, rearrangements, and reconstitutions of the seemingly solid matter, but perhaps of matter of higher viscosity under pressure and heat. It is like a mass of rolling boulders coming down a hill, being ground up and reconstituted with, say, hardened mud. Or it may be likened to a wide pan of boiling water over a small flame under the center, wherein the boiling substance rises bubbling at the center of the pot, flows up on the surface and toward the rim, and dives to the bottom and inward along the bottom cyclically as it keeps building more heat. When the source of heat is at a peripheral ring of the pan, the circulating process may be in reverse.

In addition, an inward-outward cycle of matter within each layer of the Earth's shell in a slow period, but differentially for each of the solid layers of continental plates, seems perpetually ongoing. "The state of all things is in flux."

The liquid parts of the Earth participate in the slow inward-outward cycle of rollovers by serving in absorbing some of the shock and friction of the less pliable matter breaking up and reconstituting. But stagnant liquids keep rolling over until absorbed, or evicted through the mass to the outer surface. The more solid and dense parts tend to break up to perform more localized inward and outward turnovers while slowly re-solidifying. Within each layer of the Earth's shell, there are localized rollovers alternating, turning outward away from the shell's equator on the outer surface, and turning inward in various stages toward the inner equator. (See figure 21.2, a single layer exemplarily shown.)

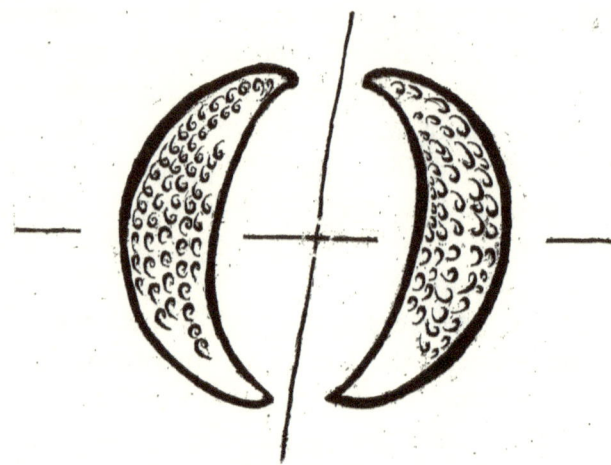

21.2 — In each layer of Earth's shell, the internal localized rollovers are outward from the outer equator, and inward toward the inner equator.

Localized turnovers are actually observed in the hard, rocky mass at the bottom of the oceans. Whether these turnovers are invasive of other layers and whether they roll opposite to each other, or oppositely in each other layer, or rub by each other.

However, this phenomenon shows that the Earth is not glued together, as by construction of masonry, mortar and rivets, not of material steadfastness, but by an agglutinative web of electric forces and energies, akin to what keeps a tiny atom active and from collapsing.

The inward-outward rollings of the plates, with the slippages between them, and the localized rollovers —besides spewing volcanic ash— generate, produce, alter, synthesize and grow the minerals, stones, lodes, fuels, mysterious edibles, etcetera, and push all such toward the Earth's surface, while some erupt to exposure (as pimples on a person's skin). Where in one century, there are the flourishing mines, in the next, new ones are discovered elsewhere, not previously detectable.

As the eccentricity of the orbital path tends slowly to approach unity, the ellipse slowly transforms to a circle. While doing so, it strives to maintain the same orbital area with the Sun at the center of the circle, per Kepler's law. (See figure 21.3, plan.)

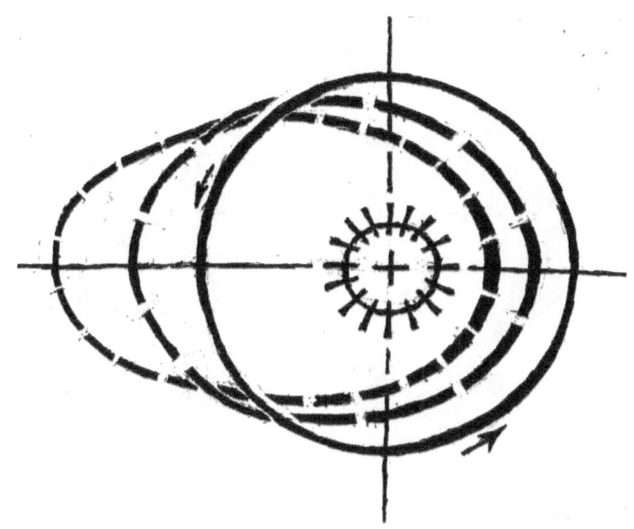

21.3 — Ellipse returns to be a circle centered on the Sun.

A perfect orbital circle would bespeak a perfectly vertical axis of rotation and the merger of the orbital and equatorial planes, going though both the Sun and the star Seirios.

The northward thrust of the continents and the outward-inward rolling of matter at the poles, as is happening in our era, will be halted, and the localized underground rolling will increase in size until the Earth is divided into a northern and southern half of rolling (which is imaginatively theoretic), wherein both poles equally will be swallowing the Earth's continents, while the equatorial zones more dexterously will be vomiting out new lands and waters from underneath the continental plates. There may arise or take form in the stead of the present continents, as we know them today, two new continents, a renewed Pangaea and a new Anti-Pangaea, each respectively over each pole, while a volcanic-like belt, whether as an exposed ridge or underwater, would possibly straddle the equator. (See figure 21.4, plan.)

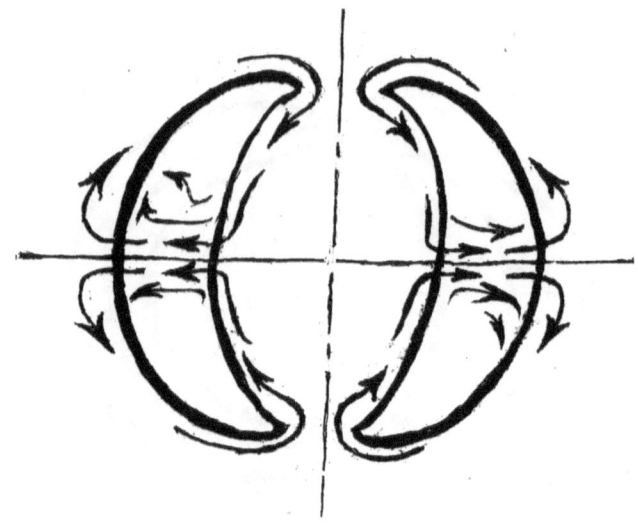

21.4 — Watery inflow at both polar openings is equalized, outflows seep and erupt through Earth's shell at belt, at greatest centrifugal force.

An active continental demarcation in our era, however, between north and south is presently prevented due to the precession of the axis of rotation externally and the askew centripetal forces internally, as well as by the equator's cyclically shifting north and south of the orbital plane, all which, however, serve to agglutinate the continents more securely (as if sewing them together).

— — —— —

Upon a chance further elongation of the orbital ellipse, the northward thrust of the continents would increase and the localized terrestrial turnovers would decrease, but would usher in transitional chaos and increase the magnitude of the turnovers. The internal terrestrial heat, mostly at the equatorial region would be intensified, for the cyclical undulation above and below the equatorial plane, while trying to maintain gyroscopic alignment with the orbital plane, thereby aiding the convoluting processes.

— — —— —

22 — EARTH'S COUNTENANCE FLEXILE

The causes and effects upon the Earth's countenance are infinite. As the angle of the Earth's axis of rotation with respect to the orbital plane changes, it is possible that the Earth's axis of rotation at times comes toward direct perpendicularity, ninety degrees, to the orbital plane. The ellipticity of the orbital path approaches a circle, with the Sun resting in the center. The orbital plane and equatorial plane merge into a single plane, leaving a more perfectly delineated and stable demarcation between the Northern and Southern Hemispheres. Then the length of day and of night becomes equal throughout the year. The Earth's present oblate (tangerine shape) sphere approaches a more perfect sphere. The seasonal changes even out to naught, zero eccentricity, uniformly in both hemispheres. The equatorial zone bears forth a Meridian Ocean with stronger violent activity emerging in the form of volcanoes and gushers, and an appearance that the insides of the Earth is in the process of turning outward.

Stretching the imagination, should the Earth's axis of rotation ever come to incline down to lie flat on the orbital plane, say, at some future age, but keeping to a given cosmic direction, the Sun would pass once yearly past each of the two poles, marking the solstices with the Sun's rays shining through the polar oculi. That is, a beam of light would shine forth from the pole on the dark half in the yearly cycle, mindful of the aurora borealis, and shine through the other Pole. The Earth's sphere would develop a valley around the equator, approaching a dumbbell configuration, to gyroscopically facilitate stabilization from wobbling of the axis of rotation.

In the case of the Earth's axis of rotation lying down on its orbital plane, while fixedly facing the Sun, the Sun would always shine through both polar oculi. One only needs to watch a child's spinning top, the swirling changes in its precession as it decelerates, until falling to its side, rolling around for a spell while pointing to the original focus, until dissipating the last of its energy.

Assuming the magnitude of the ellipse is not fixed but periodically enlarging and reducing as well as changing in ellipticity, eccentricity, should the velocity of the Earth's travel and its rate of rotation not change, then the number of days to the year is altered. The Earth's orbital plane wobbles rhythmically and twirls about the Sun in every direction, relative to the constellations. This alters, helps, or encumbers the planetary and astral influences upon each of the elements and organisms in the life of the Earth.

Again assuming the axis of rotation lies flat on the orbital plane, but keeps focused upon the Sun, the one hemisphere, always faces the Sun, a most pleasant land, and the other never sees the Sun, a most uninteresting land.

— —— —

23 — BREAKUP OF PANGAEA

The idea that at a primordial time, there was only one simple continent, perhaps floating, called Pangaea, and that it broke up and its sections floated apart and away from each other, based on the givens herein, is a rather acceptable theory. Where is this original single continent located? Some speculate in the middle of the Pacific Ocean, using the volcanic ring around the Pacific Rim as evidence. Some say that Pangaea is the presently capped South Pole, and that it is the leftover of a much greater Pangaea, where from her edges chunks of land had broken off and sailed northward to become all the continents.

Pangaea may be imagined as a warm polar cap instead of the icy Antarctica of today. The lack of inclination to the Earth's axis of rotation keeps both poles perennially in summer, in full daylight and free from wobbling. If so, would this bespeak another, an anti-Pangaea straddling the other pole.

Another explanation: Were the Earth to be revolving about its axis more slowly, possibly the Earth's axis consistently would lean toward the Sun at its southern pole, and consistently away from the Sun at its northern. That is, a full precessional cycle of orbital travel would define a cone with base projected into the southern Heavens. Thus, the North Pole is eternally in darkness, and the South Pole lives in eternal morning. Then the possibility for existence of only a single continental Pangaea would be more likely. (See figure 23.1, Earth portrayed as a spinning toy top, assume south upward.)

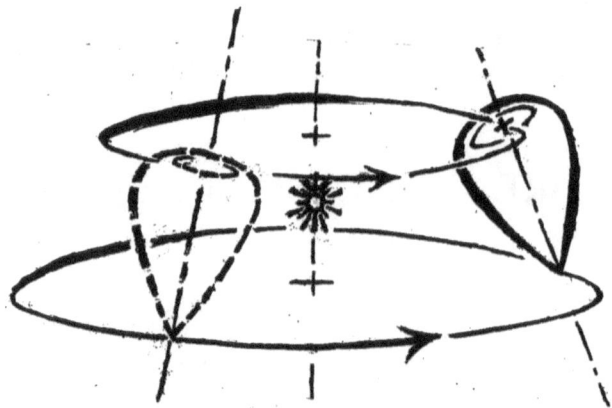

23.1 — Earth, as a decelerating spinning top, with axis of rotation in precession over the millennia leaning inward, points to a single star.

Should Pangaea have slipped sideways, perhaps from a whiplash at a newly developing perihelion, instead of uniformly northward, it would have caused some wobble in the off balance rotation of the Earth. The cause for the slipping could be internal, celestial, or both. Pangaea, then trying to move northward and stretching apart laterally, is bound to break up because of the differentiating or widening curvature of the Earth. Chunks of Pangaea begin to traverse the widening Earth's surface, and because the forces, especially the centrifugal at the points farthest away from the axis of rotation, had begun to tear Pangaea further apart. Simply, Pangaea had burst at the seams as she had begun moving laterally away from straddling the South Pole, moving northward toward the equator (of a fattening world).

Onward north, the continental plates continue to be torn apart as they move away from each other while still within the Southern Hemisphere. And in the widening oceans the pieces of plates tend incrementally to rotate clockwise in the Southern Hemisphere, per the laws of physics, decelerating in velocity of twirl until crossing the equator, when they begin rotating counterclockwise and so accelerate while moving farther north.

After crossing the equator northward and moving northward to the narrowing oceans, the plates begin to pile up and crush upon each other, with shearing and folding actions, as the field of mobility is being constricted, and each continent begins to reverse its rotation to counterclockwise, along with the waters likewise reversing their swirl. The stresses cause the rising of mountain folds. Continuing northward, the landmasses begin crowding each other, tangentially exerting the smaller of the islands to reverse their turning upon their gear-like contact with each other.

In Greek mythology, the land mass was singular and was surrounded by Ocean. And according to Genesis, four rivers sprang forth from a center source, allegedly in Pangaea. At that time, there was no rain or snow, no summer or winter. Then when the flood came, both rain and waters, "springing from underneath", had engulfed the face of the Earth. This story certainly raises the possibility of Pangaea breaking up, with segments thereof floating away from an original position straddling a pole. And some plates being stretched, rather then breaking apart, submerge. The Earth's rotation originally was upright to its orbital plane, and suddenly it tilted — shocked off balance?

The "four rivers," rather than any other odd number, is significant in the study of mechanics. Any slight increase of rotational speed, when of an even number of divisions, would set forth a vibration and a resulting destructive wobbling (as noticed on ceiling fans with four blades, instead of three or five, and in the odd number of five bolts holding in place each hub cap of automobile wheels to prevent excessive vibrations). Hence, as expected, there comes the disordering of the continents, into their splitting into smaller continents, or large islands, as they break forth out of Pangaea.

Ever-approaching closer to the North Pole, the chunks of plates are forced into fusion as each circumferential parallel progressively shrinks, to form new larger continents. In reverse progress, the tendency is for all the continents to fuse into a new Anti-Pangaea. The process of continental fusions could be accelerated, if the Earth's axis of rotation increases, or the process of fusion may be arrested by the angle of precession approaching to upright — a problem for an astronomic mathematician.

— —— —

24 — EARTH'S MAGNETIC AXIS

The magnetic axis of the Earth lies askew to the axis of rotation, without tangency thereto, and it is not permanently fixed in relation to any location of the Earth's center or outer crust (mindful of the conditions represented in figures 11.7 and 11.8), yet it defies any solution. In our times, the magnetic North Pole wanders about in northern Canada, and the magnetic South Pole wanders about in the part of Antarctica off the sea between New Zealand and Australia; and the distance of the magnetic poles are in unequal distances from their counterpart geographic poles. The angle of the magnetic axis vacillates between 180 to 120 degrees longitudinally. The universal distribution of the watery concentrations and landmasses, and other germane motions, are unequal. The magnetic axis today unexplainably lies under the Pacific Ocean, away from the average of the landmasses.

The magnetic axis, interacting while traveling around the Sun, also generates electric charges, which contribute to other repercussions. The revolving of the Earth must be affected by this, it being reasonable to suspect that the rotational speed of the daily cycle is uneven, with a decelerating half and an accelerating half (in spite of the electric clock).

It appears the Pacific Ocean fills a difficult-to-explain void in the Earth's outer crust. Water rushing in from the other half of the Earth, the Earth perhaps found a new centroidal axis of rotation, and the balance of continents must have been affected. Perhaps the magnetic axis is a wobbling remnant of the axis of rotation of yore, before the coming to be of the Pacific Ocean. Or perhaps, it is in alignment with some other celestial body.

Another possibility for influencing the magnetic poles and magnetic axis is the electric arcing of an electric current passing from the deeper part of the irregular edge of the one polar oculus, through the Earth's hollow, to the deeper part of the edge of the opposite polar oculus, the one pole positive, and the other negative. Since the poles are openings, the arcking strikes between the geologically most suitable points on the circumference of each of the two openings, overcoming or bypassing obstacles in the Earth's hollow.

As the polar oculi constantly change shape, causing the arcking points to pull every which way, the electric bridging keeps seeking an alternate set of points. The magnetic poles periodically shift, at no identifiable rhythm or reason. (See figure 24.1.)

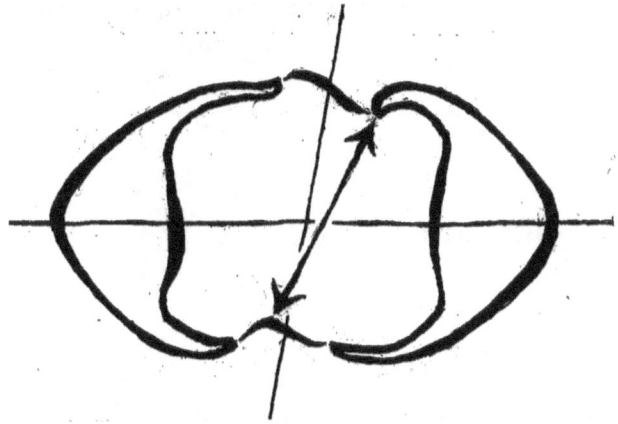

24.1 — The shortest distance between irregular edgings of north and south dimpled polar openings is ideal for an electric jump.

Assuming the Earth's magnetism is also under extraterrestrial influence, the askew position of the magnetic axis may cause a sine wave or helical path in the Earth's orbital path. And to what path among the stars does the line of the magnetic axis with its nutations trace? This motion may be independent of the earth-moon barycenter and may combine to form a more complex path.

— ———— —

EARTH'S
PLIABILITY

Natural flexibility and versatility maintain good balance with cyclical stresses and strains being absorbed by the earth.

25 — MOON BORN OF WHAT

Some geographers suggest that at a primeval time, a large continent might have peeled away from the Earth, leaving behind the void to be filled with the surrounding Pacific Ocean. The escaping terrestrial peel allegedly curled up inside out to form the planetary satellite Moon. So, then, is the hollow inside of the Moon lush with greenery, lively waters, and with higher forms of life?

From Mythology, some say, the planet Moon jumped orbit from the star Seirios, or from elsewhere, and remains captured into the Earth's orbit. If so, this would cause a drag on the Earth's orbit, which would add more days to the older ancient year of 360 days, as circumstantially documented.

The magnitude and eccentricity of the Earth's orbit about the Sun, and the particular position of the orbital plane relative to the constellations, are likely contributory to controlling the distance of the Moon from the Earth.

Allegedly, the Moon, being a satellite of the Earth traveling around the Earth, not circularly but elliptically with an elastically varying distance from the Earth, may at some time be drawn back to be recombined with the Earth's mass or fly off like a missile to relate to another planet or galaxy. Or, another continent (a "flying carpet") again may peel away at any time to form a sister Moon. Or should the Moon be set into rotation, it may fly off to become an independent planet rather than remain a satellite.

Also, the distance of the Moon from the Earth, dancing toward and away, might be determined not necessarily from the polar axis of the Earth but perhaps quite realistically from its present magnetic axis, a historic central axis, or some other compromising point between them.

But for a satellite such as the Moon, or several moons, to fly off parabolically and depart from the Earth's direct influence would cause a reaction upon the Earth —perhaps it had so happened— with a note of destruction or near disintegration. But as for a neater peeling away of a portion of the Earth's crust to seek a specific distance from the Earth to form the Moon, there had to be a critical alignment of nodes in the governing cycles and position of the Earth and Heavens to cause the necessary decisive whip lashing shock and a stop in space; or to find, or amalgamate into a needed new stabilizing rhythmic equilibrium. Perhaps the barycenter arm between the Moon and Earth may be essential to perpetuate life on Earth and, perhaps, on the Moon.

The rotation of the Moon relates to the Sun, with 365 and a fraction rotations in each Earthly year, synchronically with the Earth's orbit. That is, from the Earth only the one side of the Moon is always seen. Pirouetting about the Earth, it wobbles a little to the left and to the right, and a little up and down, so that from the Earth the three-fifths of the Moon sequentially becomes visible to Man on Earth.

Some authorities say the Moon is gradually pulling farther away from the Earth. But is this part of a cycle yet to be defined?

Of course, the distribution of the lunar mass cannot be internally symmetric, as the Moon does not rotate relative to the Earth. Because of centrifugal force, the greater part of the lunar mass must be in the part always farthest away from the Earth. As expected, there may not be any distinctively defined polar openings, for the lack of any substantial axial spinning. The Moon in its equatorial plan section is an ovate with its rounder face permanently attracted to the Earth, and its flatter rear never faces the Earth, as viewed by satellite. (See figure 25.1, ovate in vertical section, vertical axis is the rocking axis, arrow pointing to Earth.)

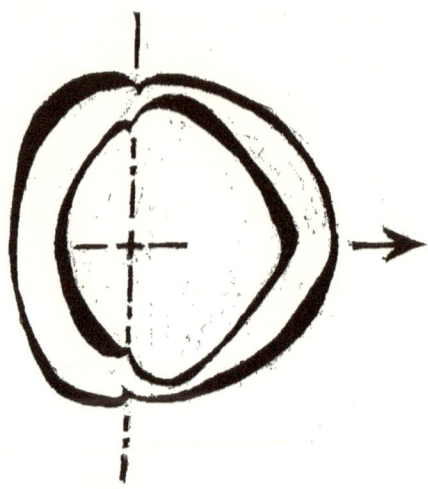

25.1 — Moon in section, depicts a configuration derived from rocking rather than from rotating, as for distribution of its mass.

The ovate shape depicts the possible scheme of solidification of the rocking mass. Near polar openings, or spindle openings, or relief openings, are not shown but assumed to exist to perpetuate the Moon's orbital position in space.

Much is speculatively said, on which the undersigned is not prepared to comment further, but welcomes further thoughts, possibilities, and information.

— ——— —

26 — SUN EARTH MOON MOTIONS

The Earth, Sun, and Moon act interdependently in their motions, not to exclude the balance of the entire solar system. Relative to the Earth's traveling along its orbital path about the Sun, the Earth is subjected to several periods of contemporaneous movements. The definition of each such period ranges from any originating motions to reactive motions and to compensating wobbling, as in harmonic overtones in the physics of music.

The respective diameters are: Sun 864,000 miles, Earth 7,926.4, and Moon 2,169 miles. The diameter of the Earth's protective exospheric limits vacillates at more or less at 800,000 miles. The diameter of the Earth through the poles at sea level vacillates at more or less 7,900.

The axis of rotation of the Earth is perpendicular to the orbital plane, which extended, goes through the Sun. The angle of the axis of rotation (tilt) is perpendicular to the equatorial plane, which extended, gores through the star Seirios. The angle of the precession vacillates off verticality at about 23.5 degrees, and so the angle between the orbital plane and the equatorial plane is 23.5. (In 1910 the angle between the two planes 20.3 degrees.) The limits of each of the two tropics straddling the equator are at 23.5 degrees latitude. And the two polar circles likewise, each is at 23.5 degrees from its Pole.

— —— —

The Earth and Moon together with their common centroid, the barycenter, relate to the Sun as a single mass. The barycentric arm to the Earth is quite short, and the barycentric arm to the Moon is quite long, which balances out their moments of inertia. The barycenter lies 1,000 miles within the Earth, as the mass of the Earth is roughly eighty-two times greater than the Moon's. The Moon's orbit around the Earth is elliptic, while the Earth serves as the base focus to the lunar orbit. The length of each of the two arms is elastically variable to absorb the shock from the compound of both the Moon's and the Earth's variables of orbiting, the conditions of which are rarely repetitive. The Moon throttles the Earth's internal wobbling into stability.

The barycenter travels relatively smoothly on the orbital path around the Sun, rather than the Earth's center. The Earth and Moon each move helically but irregularly so, opposing each other, winding above and below the orbital plane.

The Earth, because of its greater mass, barely traces a helical path, while the Moon has a more fully defined helical path as it ostensibly revolves in a more sensitively complicated manner about the orbiting Earth. The line of the two barycentric arms turns about the barycenter in almost every direction but short of parallelity with the Earth's axis of rotation. Due to the scalar difference, the Moon is orbiting quite irregularly, while the Earth's orbiting is faintly regular. With each cycle of the barycentric line about the orbital path, the forward motion of the barycenter in its orbit about the Sun is subjected to surging. (See figure 26.1, view from the Heavens.)

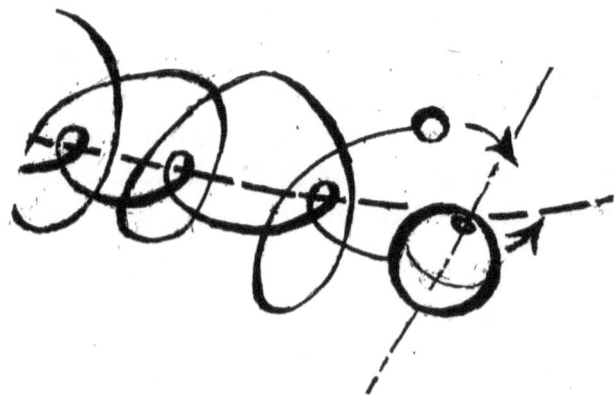

26.1 — Motions of the Earth and Moon appear arrhythmic and jerky along the orbital path around the Sun.

If, in a mathematical analysis, while on its orbit around the Earth, the Moon bounces and wobbles more than expected, by cursory calculation pirouetting nearly every which way, it may indicate the Moon must be hollow and flexile, but if less than expected, the Moon must be more rigid, but not by any means solid.

Each of the two helixes, Earth's and the Moon's, rotates about the barycenter at a timing seemingly lagging, surging, and perhaps bending the barycenter. The elasticity in the length of the line of the barycentric arms elastically varies. Perhaps, independently of each other, they absorb the rebounds for short durations. The orbital path of the barycenter also may not be along a perfect ovular path around the Sun. (It is as though two unequal in size balloons are innocently and sensitively bouncing around without incurring damage.)

Whenever the barycentric line happens to point directly to the Sun, at that moment, there is an eclipse, either of the Moon or of the Sun; and if pointing exactly to the center of the Sun, it would cause a totally perfect eclipse.

— ———— —

27 — EARTH'S PEAR SHAPE

In the present era, the continental landmasses dominate mostly the Northern Hemisphere. In balancing compensation, over the South Pole there accumulates a considerable mass, like a counterweight.

Antarctica's mountainous mass may be floating, but if so, it is trapped into position by hydrodynamic forces, or it might be frozen into position but discontinuously so, avoiding to seal the oculus all around. It may be composed chiefly of ice with earthly matter suspended within, or it may be earthly matter with a thick coat of accumulated ice, as a single mass wobbling over and covering the southern oculus. The icy accumulations at the North Pole are comparatively of very little mass, while other conditions are similar as at the South Pole.

With the continental masses, including the Antarctic, being partially above sea level, the water mass, covering the Earth throughout, tends to be redistributed throughout the globe in compensation to seek sphericity, so that the Earth may spin with lesser turbulence. The Chandler wobble perhaps supports such peaking of the waters; rather, it buffers the pounding waters from causing shore damages.

The waters around the equator and in the southern latitudes peak higher in around the Earth to balance out the continental masses above sea level of the Northern Hemisphere. The ocean surfaces, then, do not define a symmetrically smooth spheroid but rather a pear shape, in vertical section, with the wider part being the Southern Hemisphere. (See figure 27.1.)

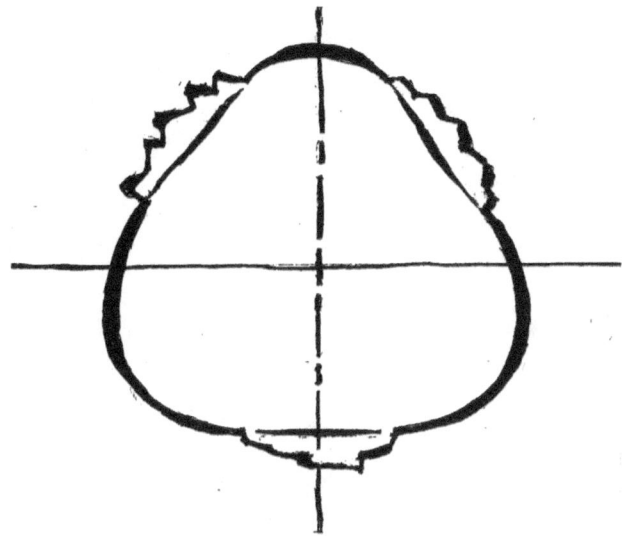

27.1 — Pear-shaped Earth exposes lands in Northern Hemisphere and Antarctica, while oceans peak in Southern Hemisphere and equator.

The pear shape is a function of differentiating balances within the masses, perhaps affecting the outer surface more than the inner surface of the Earth's shell.

Assume that in some past times, the polar caps are cumulatively overly massive, and assume the Earth's landmasses are more evenly distributed around the globe. As the Earth keeps to its stability while rotating, it has to expand at the equator and contract with respect to both overly massive poles. For the Earth, then, bulges out at its equator and dimples into the poles (flywheel principle).

The shape defined is an irregular oblate spheroid with a more pronounced circumferential rise and with an icy dome built up over each of the poles, as it would appear from outer space near the orbiting plane. That is, the globe would be ringed with two subtle valleys in its configuration over land and water, one valley in each of the two mid-latitudes, or tropics, about equidistant from the equator. (See figure 27.2, a distortion from a hypothetically perfect sphere in dash line; arrows show the two longitudinal rings of valleys.)

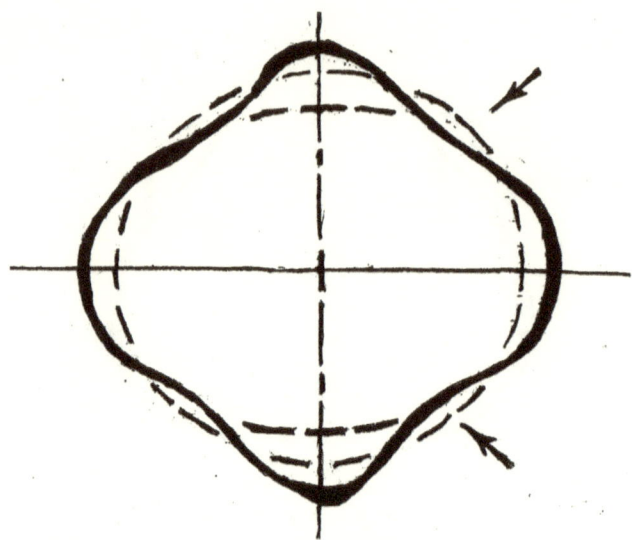

27.2 — Polar regions equally massive with ice caps on an oblate sphere are balanced by higher levels of water at equator and two tropic valleys.

Otherwise, should the Earth be a rigid sphere, whether solid or hollow, with the massive accumulation of polar ice caps, there would be an immanency to topple or rip open.

Accepting the Earth as slightly pear-shaped, relative to its present land and ocean surfaces, and assume both polar caps become disproportionately and increasingly massive for the Earth to keep its stability, it further enlarges at the equator and squats deeper at the poles and the two valleys deepen farther, one in each of the mid-latitudes. In the new shape, approaching a bloated discus, there may occur two valleys or a compounded valley in the northern mid-latitudes, caused by the distending of the landmasses and a single valley in the bulging southern mid-latitudes. (See figure 27.3, inner surface outlined; arrows show outer valleys.)

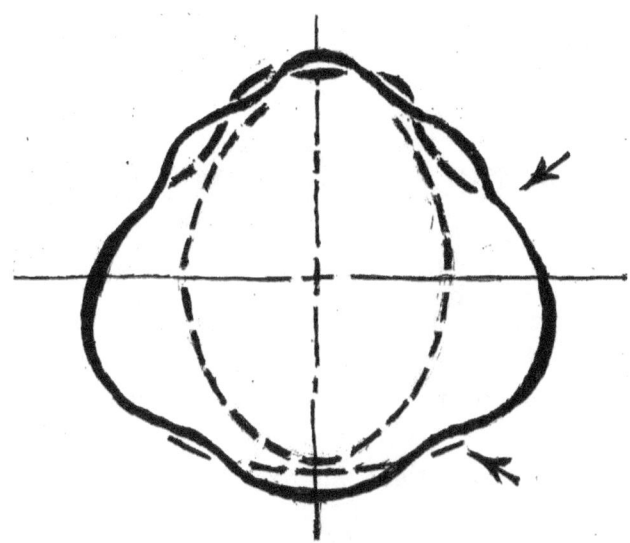

27.3 — Polar regions loaded with ice on a pear-shaped Earth yield a double circumferential valley and floods in the Northern Hemisphere.

It is speculative how much of the entire crust of the Earth, the more rigid part of the shell of the Earth, is influenced into assuming a pear figure, or is the distortion primarily taken up in the watery coat?

Were the Earth to be inflexibly solid, it would not be able to accept any deformation to absorb the otherwise most destructive variety of shock waves.

— ——— —

The varying conditions at the Earth's polar openings, the floating islands, ice mountains, slipping plates, opening and shuttering oculi, etcetera, serve as a throttling of the Earthly kite's soaring in relation to other changing celestial conditions. The changes are manifested from either orbital, cosmic, or other changes, or, oppositely, they cause compensating reactions upon the orbital and cosmic conditions. More likely, these are of mutual balancing reactions soundly to maintain the cosmic government.

— ——— —

28 — CHANDLER WOBBLE

Assuming with each cycle at perihelion, the Earth is subjected to a sudden upward displacement, while at the same time it remains in its pear shape, something would happen similarly to the Chandler wobble.

With each fifteen lunations, approximately, the Earth undergoes a subtle cycle of undulation of spheroid flattening with rotational deceleration and return. Each cycle of undulation affects the northern and southern hemisphere untimely, arrhythmically contributing to a torsional oscillation in the globe. The observed phenomenon is known as the Chandler wobble.

If it can be established that about each year and a quarter there is such sudden upward thrust and gradually back downward, the Chandler wobble might come to be explainable. The prevalence of the northwest-southeast direction in the formation of most of the Earth's mountain ranges bespeaks a deep geologic torsional distortion.

With each undulation taking place, while the unending inside-out localized rotations are in progress. The solid, liquid, and gaseous matter in each of the layers of the Earth's shell are squashed and rearranged, as through a kneading action as well as a pumping action, which yield water springs, volcanic eruptions, gas and petroleum flows, and particles of radiation.

Each motion or wobble, each change of direction and speed of travel, each change in the differential rate of rotation, each flattening or rounding of the sphere, and each approach or passing nearby of other heavenly bodies, and each rebound thereof affects the configuration of the Earth. Affected as well are the Earth's oceans and the flow of waters, the biologic and chemic performances of flora and fauna, and indeed the intricate calendric concept of intermeshing cycles of time for Man's bountiful activities.

Each form of changing motion affects differentially the varying densities, pressures, durometer of materials, gravitational intensities, adhesiveness, etcetera, of all matter under and over the surface of the Earth. Such are reflected in tides, floods, humidity, temperature, earthquakes, submergence and emergence of the landmasses, configurations of oceans and continents, flora, fauna, colors, rhythms, and many other terrestrial phenomena. Such motions, wobbles, etcetera, of the Earth can also cause the centrifugally spinning off of moons and debris, which would leave a change of weights, momentums, moments of inertias, rotational timing, and the rhythm of all motions of the Earth. As well such reflect in the adjustments in the Moon's characteristics, as truly bound to the Earth.

As the size of the Earth's orbital ellipse increases, the speed of travel increases, the pull of gravity increases, solids are compressed, the size of plant and animal life shrinks, the surface area of the oceans increases, the Earth's spinning about its axis decreases. And as the size of the Earth's orbital ellipse decreases, all the activities of Chandler wobble begin to mitigate.

— ——— —

29 — WIND AND WATER CURRENTS

The Earth's wind and ocean currents are motions associated with the friction in the rotational drag through the atmosphere, to which contribute the exospheric and intra-terrestrial factors. This is evidenced by the directions of the Earth's mountains, valleys, watery channels, passages between bodies of water, underwater mountains and valleys. As directional vanes these influence the winds and currents.

The waters generally meander toward the poles and ultimately flow into the poles to disappear under the ice caps and into the polar oculi. Some follow subterranean passages or flow upon the inside surface of the Earth's shell toward the interior equatorial regions, and then distilled, they find their way outward through the shell by centrifugal force.

Do the waters reenter the oceans, say, by seepage outward around the middle or equatorial latitudes? Are the waters within the Earth broken down into hydrogen and oxygen to erupt violently through the Earth's crust, as in volcanoes, or to seep smoothly, like inconspicuous gasses being breathed out? Does such heat and churning serve to manufacture gasses, petroleum, and fresh spring water as well?

Winds from high altitudes drop down toward the Earth's equator, pick up heat and absorb humidity as they generally flow toward the polar regions. Then they rise to a high altitude losing their humidity and heat for their return trip. This is nature's way of preventing disuse of the equatorial regions.

The more perfectly spheric the Earth is, the gentler all these currents are. The winds and waters would divide apart in both directions from the equator.

The more oblate the Earth is, the more briskly the currents are tossed and become parted toward the poles, subjecting the equatorial regions to a drier climate. Circumferential valleys and mountain ranges in the configuration of the Earth create a haven against winds and induce a more abundant but gentle rainfall. But they generate more turbulence in the waterways and sea straits (see figures 23.2, 23.3).

The changes in the speed of rotation of the Earth affect the course of the currents, of winds and waters, and in their changing intensities, volumes and temperatures.

Water vapors either settle down cyclically with each sunrise and sundown generally. Or they collect into clouds for indefinite periods of time to be carried by the winds to be disbursed through quantities and kinds of electric charges into rainfall and snowfall. All in turn are dependent on the Earth's configuration, activities, and repercussions, as though by actions of a dynamo. The migrating flights of birds and paths of fish, which are not always in synchronicity with the winds and currents, may reveal more secrets.

It is possible that at one time, there were no clouds or rainstorms on the Earth, according to ancient stories. Should the Earth come back to a more perfect sphere, all should again be according to the ancient stories.

— ——— —

30 — SHELL GENERATING HEAT

Besides the heat received from the Sun, the Earth generates its own heat. Among the several sources of generating heat in the hollow of the Earth, there is the whole of the Earth's shell itself, with its frictional stresses, strains, pressures and chemically generated forces.

The centripetal and centrifugal forces acting on the shell's mass are equalized at about the shell's median depth. As the mean depth of the shell is greatest under the equator, and diminishing toward the poles (shell in section being crescent-shaped), likewise the two opposite forces of equalizing compression are greatest under the equatorial regions, diminish laterally to approach zero at the poles. With compression producing heat, the greatest amount of heat within the mass of the shell occurs in a belt-like area midway under the outer equator and over the inner equator, with heat diminishing laterally in both directions toward the oculi.

Per the Kirchhoff theory, with a heated equator drawing ions, a temperature drop is caused at the ends of the Earth, from the polar areas. The near polar regions are left with permafrost, which begins next to the shell's oculi, spherically permeating away from the vicinity of the poles, along the mean depths of the shell, while dissipating generally before reaching the tropics. The entrapped massive caps over the two polar oculi somehow build up ice from underneath, as the waters flow by and under, besides a little accumulating from precipitation.

With a decreased speed of rotation, the heat belt under the equator slightly cools for the lessened compression, and in compensation, the polar areas lose some of their extreme cold. Ice mountains at the poles can be set off to melt, and the equatorial zones become cooler as well as drier, while the level of the ocean differentially rises. The melting of ice, in turn, helps to restore the Earth's massive figure toward a more perfect sphere.

The polar ice caps could be diminished by other phenomena such as an increased wobble in the Earth, shaking up the caps, fracturing, or inducing friction. By their being accumulated too rapidly, crushing is caused and they flow away. A change in the eccentricity of the Earth's orbital path around the Sun induces stress and temperature differences.

The lengthening of the orbital ellipse (increasing its eccentricity) generates more frictional heat in general, and yields greater extremes between summers and winters.

Nothing is stable, and more causes and effects are yet to be discovered regarding the activity of planet Earth in its galaxy to its staying functionally alive and well.

— —— —

EARTH'S
ENVIRONMENT

Protective bubble encases the Earth, allowing the intake of
filtered celestial nourishment and screening out toxicity.

31 — EARTH'S ATMOSPHERE

The atmospheric and higher exospheric gaseous layers protecting the
Earth from physical clashes, chemic poisoning, and deadly radiation is
a complex matter of briefly "bulletproofing", "cushioning", repelling,
and insulating the encapsulated Earth. The Earth's atmosphere is
composed of several layers protecting and conditioning the Earth's
surface and its entirety. Each of the atmospheric layers successively
outward becomes more exaggerated generally from flatter tangerine
to rounder apple shape, each thicker at its equator and with more
deeply pronounced dimples over the polar regions.

Each of the atmospheric layers rotates with the Earth, but each
layer successively outward rotates more slowly, due to atmospheric
drag and the circumferential increase, requiring greater velocity to
keep up. The shape of each layer is in incessant undulation, both
in the spheric and torsional modes, and it is subjected to distortions
from wobbling. Each layer generates beneficial activities of electric
charges, eddy currents, capacitances, and "short-circuiting" internally
and in conjunction with its neighboring layers. Each layer undergoes
a series of internal rollovers as in the Earth's shell. The atmospheric

masses with their pressures are sometimes concentrated over the equator; some are distributed more evenly over the globe; and in part, some shift with ease from one hemisphere to the other. However, they outwardly react slower to the Earth's activities. The isolating "pauses" between the layers may be considered "friction proof" and insulating zones.

The altitude of the troposphere at the poles becomes doubled during the winter months. At the equator, it is consistently double the altitude of that of the mean height near the poles. It appears that the atmospheric layers are more affected by the whiplash of the Earth passing closest by the Sun, or, one may say, the atmospheric layers are less susceptible to being centripetally drawn in by the Sun, as are the more solid of layers constituting the Earth. That is, the Earth unconditionally dominates the atmospheric and exospheric layers.

The axis of rotation of each of the several atmospheric layers lies askew to each other, and askew to the Earth's axis, and all vacillate, whether harmonically or disharmonically, relative to each other. Their relative positions are in constant change in precession, rotation, and all other motions, with axes of rotation askew to each other. (See figure 31.1, showing summer with a dash line in the Northern Hemisphere, in schematic scale.)

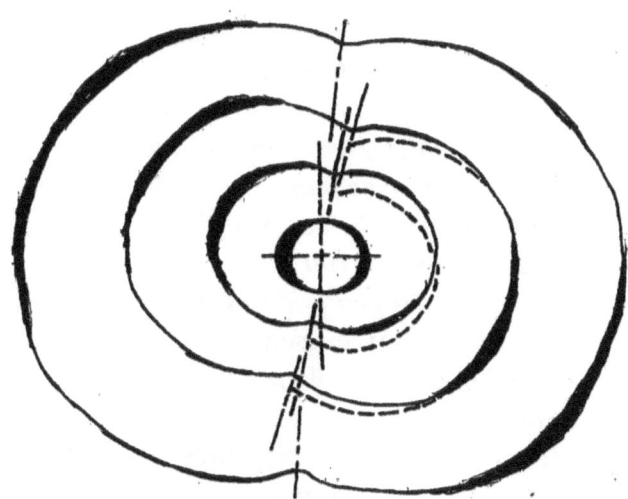

31.1 — Protective atmospheric spheric layers, each is thickest over equator, thinner at poles, thinnest at respective pole in summer.

From outer space, protons and electrons descend into the atmosphere and onto the Earth. One ton of vaporized physical matter is added to the Earth's mass each day; ice crystals also add to the Earth's daily menu.

— ——— —

Roughly, from outer space, but from within our galaxy, down to the Earth's surface, the various layers with salient characteristics and averages are:

Star Seirios, 8.7 light-years away.
The Sun, 93,000,000 miles away, average.
Satellite planet Moon,
    revolves at 238,900 miles up.
Magnetosphere, 40,000 miles up,
    with helium,
    stops certain particles and rays
    from descending any farther.
Pause.
Upper Van Allen belt, 13,000 miles up.
Lower Van Allen belt, 9,000 miles up.
Interplanetary space beyond.
Exosphere, 2,000 miles up,
    lightest part of atmosphere,
    molecules escape into space.
Magnetospheric pause.
Auroras over the polar regions, 300 miles up.
Thermosphere, 100 miles up,
    with nitrogen and oxygen,
    increase of temperature.
Ionosphere, thin, 95 miles up.
Pause.
Ultraviolet and cosmic zones, 30 miles up.
Stratosphere, 20 miles up.
Pause.
Troposphere, 10 miles up,

holding 90 percent of physical matter
suspended above the Earth.

Understanding the character of each of the atmospheric and exospheric layers —yet another speculative subject— serves suggestively to illustrate what tends to happen as well with the internal layers of the Earth.

The several layers of the human body's skin are of tissues, and muscles, which act not in synchronicity, but knead against each other to accomplish whatever. Forces through eruptions, as sweat, pimples, rashes, digestion, swelling etcetera, actually massage the body. Likewise rollovers, earthquakes, gushers, atmospheric conditions, slippages and oil, water and gas veins, etcetera, normal or abnormal, keep massaging the Earth.

— ——— —

32 — TWO VAN ALLEN BELTS

There are two major Van Allen belts —some authorities simply say three or more— girdling the Earth far above the exosphere but within the Earth's magnetosphere. Each is characterized as holding intense radiation. These are composed of sub-molecular particles trapped from outer space and charged into activation mostly by the Sun.

However, the outer of the two major belts, the greater of the two, contains helium, a basic element, a simple material substance, mindful, so to say, of a potent hard outer shell around the spinning Earth. The width of the greater belt extends beyond the projection of the mid-latitudes but comes short of the projection of the Arctic and Antarctic circles. The width of a smaller outermost belt hardly extends to the projected tropics of the Earth.

Each Van Allen belt bears its characteristic great oculus (as barrels with ends removed). Serving as lenses and reflective lenses, each appears as a near torus with the inner side scooped out and in cross section crescent-shaped.

The belts are approximately in alignment but vacillate laterally and vertically over the equator independently of each other. Each is thickest over the equator, from where the projected radiation from beyond the magnetosphere is most intense and diminishes toward naught toward the polar axis. The two belts are like electric "third rails" of the exosphere (see figure 32.1).

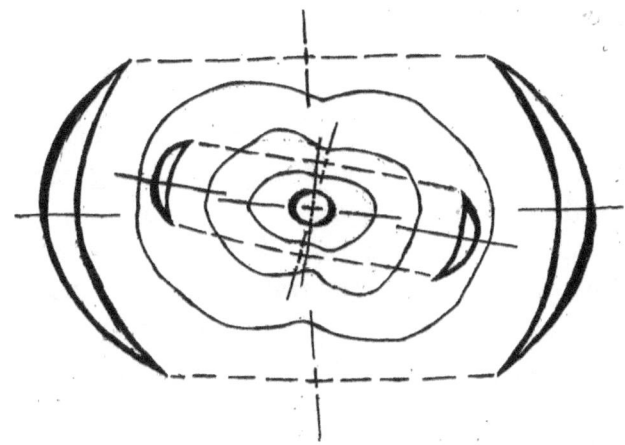

32.1 — Van Allen belts rotate askew to each other and askew to Earth's axis of rotation but keep widely clear of polar regions.

A third Van Allen belt, farther out, is thought to be of much wider extent, but very thin and like a blanket it wraps down closer to the polar regions to enclose the ends of the two major Van Allen belts.

It is reasonable to expect that each of the two belts rotates with the Earth but at a different rate than that of the Earth —the inner slower and the outer much slower— and that they oscillate up and down between the polar areas and are subjected to wobbling. The axis of rotation of each of the Van Allen belts is askew to that of the Earth's and to each other, bespeaking a disharmonic, or nodal, relationship between the two.

The ions floating within the confines of each of the two belts are in constant motion, zigzagging north and south while progressing around the belt. But do the belts bear a current that travels opposite the rotational direction of the Earth? If so, do they directly contribute to temperatures and global warming?

To be noted, vessels launched into outer space are directed through either of the two polar corridors, to avoid the friction associated with the Van Allen belts and other exospheric layers, all which bear some physical or electric resistance and danger for spaceships. However, space-vessels are not directed strictly on the polar axis, as there is something negative yet unexplainable harbored there, but are directed through the rather vast space between the polar axis and the edging limits of the Van Allen belts.

A sudden change in the slope of either of the two Van Allen belts can reveal or cut off more of the Sun, or it can change the angle of aberration of certain of the Sun's rays going through the density of the belts. Or the rays may reflect off the inside of these belts when coming from the opposite direction. Ultimately, they may focus upon the Earth. In any case, while in a transitional state, the Sun, for an hour or so, because of atmospheric aberration, may appear to stand still or move closer or farther away. The Moon's appearance, as far as texture, shape, and color, is also affected.

The best of clues for a hollow Earth, besides accepting the axiom that nature's designs endlessly repeats and reflects, are the flexile, vacillating and elementarily loaded Van Allen belts.

— ——— —

33 — RADIATION & EXOSPHERE

Each of several atmospheric layers with certain pauses of isolation, as well as of the two Van Allen layers, although intangibles, may serve to filter, refract, or absorb all sorts of solar, cosmic, celestial waves and booms, etcetera.

In accordance with the rules of optics, some rays coming from a particular direction and penetrating the atmospheric layers bend, refract, and concentrate their focus upon and into the Earth. (See figure 33.1.)

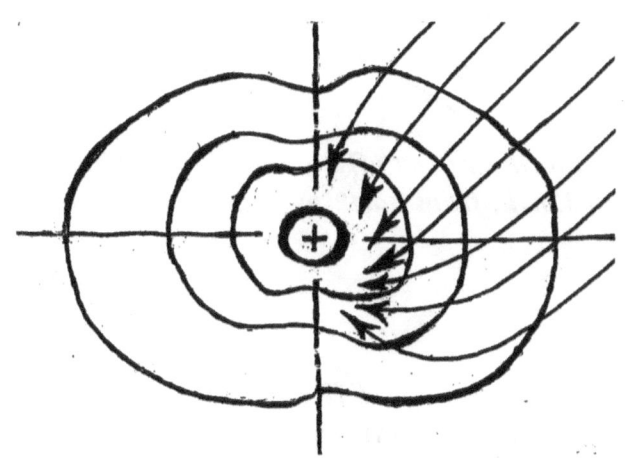

33.1 — Atmospheric layers over the Earth as lenses aberrate Sun rays, some touching the Earth's surface, some penetrating.

— ——— —

Some Sun and other celestial rays coming from a particular direction reflect off the concavity inside the Van Allen belts and concentrate their focus upon the Earth, on to either hemisphere. Hence, the more intense of rays and energy may be experienced during the winter days (as the Sun in its lower position sees the concavity more clearly), and with lighter nights experienced, on the opposite hemisphere to the Sun. (See figure 33.2, Sun rays from lower left, being bounced around as on a pool table.)

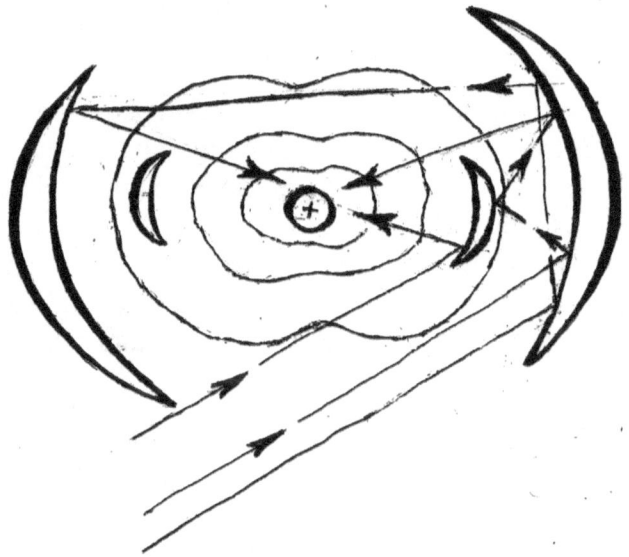

33.2 — Sun rays reflect off inner hollow of Van Allen belts and onto Earth, from north to south hemisphere and vice versa.

— — —— —

Some celestial rays of different wavelengths coming from every direction may reflect off the concavity of the Van Allen belts, refracting in several directions, some falling to, or grazing the Earth, while others clearly bypass but not without leaving an effect upon the Earth. (See figure 33.3.)

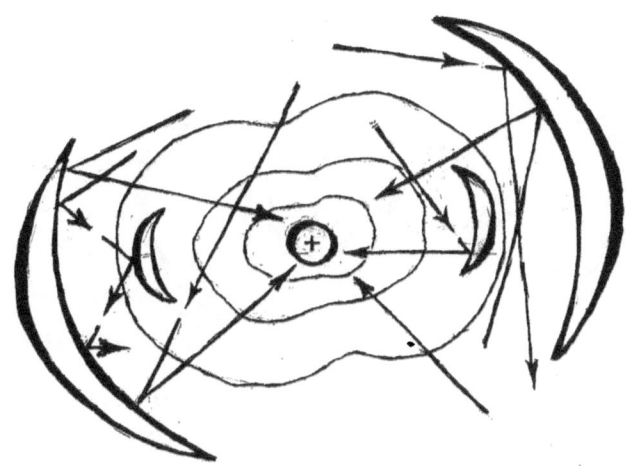

33.3 — *Random reflections of various celestial rays bounce off the Van Allen belts onto and around the Earth.*

— ——— —

The two Van Allen belts perhaps also serve to filter out certain more direct Sun rays and other rays of energy, and then focus them upon part of the Earth or through the Earth to its interior or they refract and spread across the Earth and on to the Earth's atmosphere. The filtering is strongest at the equator of each of the belts, diminishing toward the edges, where possibly the rays are needed more for the Earth's flora and fauna. Some rays are and sometimes along with an assortment of particles they carry, are trapped, converted, and stored as energy. (See figure-33.4, time of equinox.)

33.4 — Sunrays filtered out through Van Allen belts, are bent and focused or refracted upon the Earth.

Some rays are reflected back into space. Some rays mingle with physical substances in the atmosphere and are parried from settling on the Earth by the Earth's magnetic field. Are we talking about another secret of UFO's?

During the equinoxes, the solar rays are focused directly onto the equatorial region, the focal point lying under the equator inside the Earth. Yet the equatorial regions receive sunlight of less intense energy, for the filtering is strongest over the equator because the atmospheric compound lens (lens in section) is formed of the three atmospheric layers with their pauses and the two magnetic belts, and it is thickest over the equator.

Question: Why is the equator not the coldest part of the Earth, in our age, when during the equinoxes, it receives the least amount of radiation, and during the respective mid winter and mid summer periods, the rays are focused opposite, to the winter hemisphere? When most of the Sun's rays become focused closer to the polar regions during the winters, due to the precession, why are not the winters warmer? During the summer, the mid-latitudes of the summer hemisphere are closest to the Sun, not the equator. Why, then, is the equator warmer?

In partial answer: In the upper atmosphere, in the thermosphere, there is concentration of heat, with cooling above and below. So, not much heat is directly conveyed to it, but heat is regenerated on Earth and on the lower atmosphere, from the altering of the trapped particles and incident rays.

During the mid summer period, especially when the Sun is at high noon, some of the Sun's rays enter directly through the polar voids of the Van Allen belts, striking the underside, or the inner side, of each of the two Van Allen belts from across over the Earth's poles, just as they would enter the polar area. Logically, it is expected that the summer nights would be light and the winter nights dark, for the winter hemisphere at night is tilted away from the Sun. This leaves us with the suspicion that winter nights are lighter because of more refracted and reflected light of the Sun from the inner faces of the Van Allen belts (as in figure 33.2).

Question: Why are summer nights darker than winter ones even though they are not as well eclipsed from the Sun as are the nights of winter? Why are there no nights in outer space as known on Earth? Do the rays of the Sun cause the emergence of a "darkness", or do the rays avert or dissolve the particles of "darkness", when striking the under, or inside, part of the Van Allen belts with their magnetic qualities? And, could there be such thing as a celestial night? And could such be related to black holes undulating out of hiding?

Again, in partial answer: atmospheric energies act differently under varying temperatures and from different natures of luminescence. It is assumed that the Kirchhoff theory and other intra-terrestrial phenomena have more to do with the polar cold than the alleged lack of Sun.

That is, "darkness" is a substance, not an absence of light per today's Solvey physics. Is dark matter something invisible to us human beings?

— ——— —

34 — POLAR ICE CAPS

There was a time in the long past, speculatively: The Earth's axis stands vertical to its orbital lane, that is, the orbital and the equatorial planes are one and the same. There is a rather uniform moisture content suspended in the atmosphere as well as some in the rest of the exosphere, which at one time in the past extends much higher. Such gives the Earth a greater protective shield against solar and cosmic radiation and maintains a controlled and uniform temperature nearly throughout the Earth for the optimum sustenance of life. Dew gently sets upon the Earth each morning. And water is tenuously taken up each evening.

The greenhouse effect also proffers quite a paradise to the Earth. At that time, a human being could experience a more optimum comfort of living in nature. There are no changing seasons and no drastic electrostatics at that long-ago time.

When the Earth's axis gradually or suddenly comes to a tilt, great atmospheric friction, if not the only disturbance, causes the Heavens to dissipate a great part of their stored moisture content, leaving behind the dried up Van Allen belts, if not other dried up layers as well. The Earth floods, and some of the waters begin collecting into finite and isolated clouds for much longer than a day, each time releasing the waters suddenly and at times catastrophically. Seasons come into being, and the various portions of the Earth are gradually subjected to extremes of hot and cold, wetness and dryness. See the account of Noah in the book of Genesis.

Then the excess of the waters on the Earth's surface must recede. The wobble effect in the Earth's rotating about its axis tosses the watery precipitation toward the two poles, collecting them into packs and mountains of ice, trapping the excess of waters (perhaps for their lacking a salt content) into storage, along with the suspended residue over the polar openings. Most of the excess waters leave behind somewhat deeper oceans with a balanced swelling about the equator and some collect into a continental mass over the South Pole. The major parting direction of the waters southward is opposite of the continental drift northward.

The waters react quickly to the spinning wobbles of the Earth and especially to the undulations associated with the Chandler wobble, thereby dragging more of the waters toward the rotationally dragging Southern Hemisphere.

The mounds of polar ice melt and reform with a series of ice ages, leaving us today with the frozen mass of the South Pole. The South Pole experiences longer and colder winters than the North Pole, as the Earth travels slowest past the point of its orbit farthest away from the Sun, aphelion, and therefore the waters cake up into ice more permanently than is possible at the North Pole. The whiplash of perihelion yields some frictional heat for the Northern Hemisphere.

The icing up takes place from the saltless water underneath the mass, heaving the Antarctic mass up higher, which in compensation, we may suspect, is in balancing proportion to the wake of the northward drift of the continents. Oppositely, in the North Pole, any ice caking up is constantly, yearly, subjected to breaking up into icebergs from the stress and fracturing from the whiplash effect at perihelion, as well as the ocular swallowing, while the South Pole experiences a gentler compaction in our age.

Not only the alignments of stars, our astronomer physicists may mathematically determine at how many degrees difference between perihelion and the winter solstice (as they incrementally bifurcate) would yield the most dangerous time for shattering the northern icebergs.

As the polar masses of ice are melting, and because the Earth's configuration is changing at the same time, the rise of the ocean levels may not be the same for all latitudes and longitudes. The landmasses, straits, and depths of waters and the drag of the ocean waters pushing along the western shores of all continents, serve as retention ponds. Besides the general flooding, excessively so in some regions, in compensation in certain latitudes new lands may appear to rise out of the ocean, or rather are forced to rise because of pressures being generated wherever the Earth's shell is subjected to contraction and blistering.

Assuming the Earth maintains its configuration and its rate of rotation, a change in the distribution between the waters and the solid part of the Earth's shell would take place. Generally speaking, the more firm shell would come closer to sphericity, from its present oblateness, pulling inward, lessening, the circumference of the equator, but only to be filled with the waters accumulating at the equator and its flanking latitudes. This means that the first lands to experience flooding would be the equatorial lands, possibly Congo, Amazon, and Borneo.

As the Earth yearly swings past perihelion, the increased mass of waters react more extremely as they are thrust northward for about a week to flood the temperate zone of the Northern Hemisphere, the lowlands of the Gulf of Mexico, and the northern lands of the Pacific, Atlantic, and Indian Oceans.

The increased watery surface in the warmer equatorial belt increases evaporation, but the winds carry the clouds to give rain upon the flanking temperate zones, perhaps bringing rain to all the present deserts between the temperate zones and the equator. In general, the humidity and the rainfall levels for the whole world increase, again changing the global temperature.

Assuming the Earth's shell does not change in degree of oblateness, nor the Earth's rate of spinning, but with the masses of polar ice being reduced, the waters move more violently toward the equatorial regions. Should the Earth slow its spinning, which would be the more likely scenario, the general overall oblateness would increase, the valleys of the two temperate zones would disappear lowered somewhat into the oceans, while more land would rise out of the seas in the equatorial zones. The Earth's pear shape would diminish. The wind and ocean currents would diminish, allowing an increase of rainfall in the equatorial and both adjacent zones.

The dissipating of icebergs, by a series of repercussions and reverberations, affects multifaceted ecologic changes through the entire Earth, dry ground, oceans, and atmosphere, some disadvantageously, some beneficially.

– ——— –

35 — MOLD SHAPING THE EARTH

The obvious mold would be spheric, or spheroidal, composed of gasses and particles.

The subliminal mold theoretically would be crystalline in form, composed of active and inactive energies in sublime synergy to forming the obvious mold.

As for the spheric, assume there is an infinite etheric space (an undefined space) being churned by some whirlwind, fostered by a greater government. There is great activity of invisible forces and energies of many kinds, known and unknown, being set into motion therein, some accelerating, some being crowded outward, others condensing into matter, and so on. And a stage arrives, where these stratify into several layers, each thicker over the vicinity of the perimeter (circumference) of turning and thinnest near the assumed axis, now being established. The heaviest of the layers congeals into a more rigid shell, and this we call our Earth. By this definition our earthly world is protected and adorned with quite a greatly sized environmental wrapping.

Theoretically, the energies of centrifugal and centripetal forces define the particular magnitude of our terrestrial shell, our Earth. And, other layers accordingly assume a size and similar shape within the Earth and over the Earth, in the etheric spaces. Our World may be further described as a great fruit with the Earth as its little seed. These softer shells, or terrestrial coatings, take form over the earthly shell's protection, by cushioning and containing the Earth, while being bounced around somewhere in the infinite universe.

In response to the Earth's solidification, along with its new trait of strong capacitances of kinds, somehow, the etheric layers of charges, like and unlike, identifiable and unknown, are concentrated and compelled into stratification, and the maintenance of their orbiting in various etheric positions. Each of such etheric layers coagulates into a more defined terrestrial belt. But, hardly any of the energies overreaches the Earth's polar regions.

So, with no significant pressure outward from the poles, and most of the pressure being outward around the equator, the ethero-spheric layers serve as girdling belts for the earth. Along with the protective covering belt theory, the particular material balance of the Earth is throttled; we may as well assume that through the poles there is more suction inward then of ejections, and in either case, the Earth by these also has the inbred facility for vertical movements.

This kind of spheric cover serves proportionately in harnessing the differentiating stresses and strains by the contributing of each layer into bolstering this complex mold around the Earth, inside and outside, differentially from one pole to equator and equator to the other pole.

The many spheres, as the atmospheric, exospheric, magnetospheric, and possibly others, collectively let us call "etheric spheres", or "etheric layers". As the Earth has its polar openings, each of these spheres, too, has its polar openings, irregular and out of proportion as they may be. And in cross-section, like the Earth, each of these is as though of two opposing crescent shapes, but fluffier, and more scant over the regions of the polar oculi.

The spaces between each of the spheric layers, for their non-mixing, are called "pauses", formerly conventionally considering them merely voids. These pauses interconnect wrapping around the edges of each of the etheric layers, in the series of etheric polar openings.

As each of these spheres rocks back and forth, some more relative to the Earth's equatorial plane and some more relative to the orbital plane, and yet some others to other terrestrial and celestial phenomena, their respective axes of rotation form a bundle, tightened at the Earth's center (per Figure 11-I). The nature of matter collected and disseminated by each of the spheric layers, including within the Earth's hollow, may be directly related to any one of the stars in the Heavens.

So, any major calamity befalling the Earth, upsetting its magnetic fields, or thwarting its axis of rotation, the heavenly bodies within reasonable time, would readjust, "chiropractically" back into shape. Or, they may reflect a most remote celestial calamity, by its shock "tele-energies", in reaction calling for a needed terrestrial readjustment — note, it is not astrology, another matter. (The common outlook today is that calamities are physically immediate, as by near sweeping comets.) All such etheric spheres are subject to some distortion capable of refracting and diffusing, or particularly shaping each of these layers. (See figure 35.1, imaginary zones are included.)

— ——— —

Each of these etheric spheres rotates at different speeds, relative to each other, depending primarily on their content of matter and energy, and secondarily on their distance from the Earth. So, the slippages between these etheric spheres (or girdling belts) produce different electric energies of sorts. The Eddy currents, on the one hand, by their strong repelling waves, insure the physical isolation of each shell. And, on the other hand, the shells function as a celestial electric motor, with a series of shell-like armatures. The Eddy currents also function as ball bearing spacers to prevent friction between the shells, and to prevent electrical short-circuiting. Perhaps the "etheric electrical supply" from each of the armatures (girdling shells) is channeled into the assortment of polar axes and concentrated through the Earth's open polar oculi and into its interior; and perhaps, each other etheric spheres supply, alternating, is channeled through the opposite polar oculus of the Earth.

(Imaginable? A new electric motor, learned from nature, with a spheroid armature within a spheroidal armature, and more in series within, by which multi-phase electric power is an immediate product. Frictionless antennas projected into the polar ends (oculi) of these armatures channel the power into a transformer, if not by brush-like direct contact.)

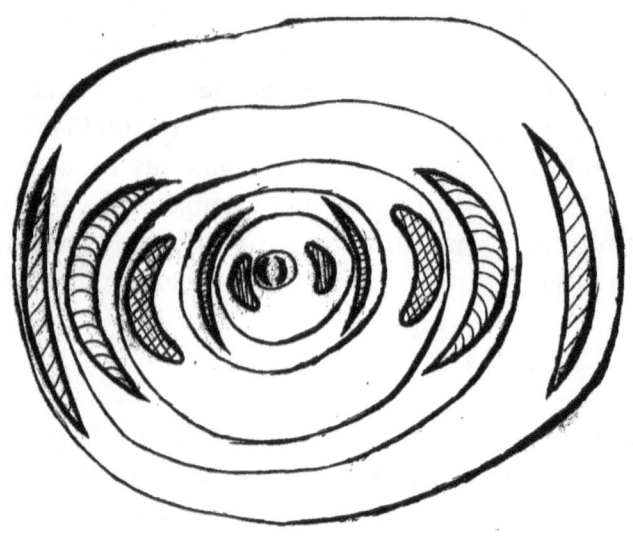

35.1 — Imaginary series of Earth's atmospheres, gaseous zones, Van Allen belts, and other matter are in successive alignments.

These happenings, deftly organized as they are, cannot be the result of a long series of chances and spontaneous revolutionary accidents, as they exhibit a high-complexity energizing potential to build and protect the planet and to form an ethereal "cement" of the galaxy.

Each of these atmospheric belts, perhaps primordially more invisible, each according to its particular charge, attracts a particular choice of celestial dust, fuses therewith and together give substance and matter into being. Other kinds of charges emerge and self-metamorphose into a more solid substance. Perhaps in many cases, the two above combine to give us our numerous elements and in turn their composites.

The tuning of our particular human senses gives us to understand axiomatically that the Earth's surface serves as the basis for dividing our abode between the exosphere with its lower layers and substances and between the terrestrial endospheric layers with certain substances. Both divisions of essences and substances are of compatibility for potentially living and dwelling therein and thereupon, whether in the forms of solids, liquids, gases, energies, organic life, or in a space not different than open to the outer atmosphere.

The unobstructed ocular polar opening to the Earth's inside is wide enough that one may not sense or visually discern his gradual entering through a polar oculus. It may be wide enough not to see across to the other side of that opening, except to sense entering into a region of a gentler climate and noticing a rather different "sky". Satellite photography shows that in approaching the region of the polar ocular area, winds and storms diminish to near calm. And yet such photo records are not published; what is the secret? Insecurity! A small airplane may enter and land on the inside, once safe navigation arts are perfected. Satellite photography shows conditions in other planets as being similar. Perhaps the momentums of spacecraft are neutralized by the siphoning qualities associated with the center line (axis) of polar oculi.

Volcanoes vomit out and meteorites deposit tons and tons of matter, some being pulverized in the high atmosphere into dust and minerals, and raining or settling these, as curses and gifts, here death and there life, upon the Earth.

In any particular sphere, or belt, or hollowed "doughnut" conducive to attracting celestial matter, if saturated, such matter agglutinates into heavier precipitation, augmenting the domain of its host, which is no less our Earth's surface.

The two invisible molds, therefore, forming the Earth's shell are, on the one hand, the centripetal near-concentric series of belts and spheres above the Earth's shell and, on the other hand, the centrifugal series of belts and spheres in the hollow underneath the Earth's shell. The mold in these cases is formed of energies.

Such successive belts of atmospheric layers are the protecting cushions. As well, such are found around each of the other rotating planets of the Heavens. That is, by mutual planetary, any celestial disaster may be prevented. Were any such invisible protective outer belt or sphere to be neutralized, the quality of molding thereof would be depleted, leaving the possibility of a planetary shell to face drastic changes, perhaps disintegration, if per chance failing in some natural compensation. It had been suggested that an orbiting asteroid between Earth and Venus was a remnant of a planet, yet nevertheless, the cosmos still survives.

A Chinese potter slaps the clay pot he is molding on his turn table to throw it slightly off center to break its esthetic monotony. A viewer walking around it may notice the lively waving shape. The atmospheric layers likewise slap the Earth around to keep it vibrant to continue "spicing" our habitat for us.

As the inner layers of the Earth are bound with sinews of energy, terulian lines and physical circulating "strings" then, too, so are the various gaseous layers above the Earth bound with comparable sinews.

— ———— —

On a greater scale, as for our solar system, we can assume that belt-like or more rounded spheres of different orders are formed around the Sun. Within certain ones of these cosmospheric belts, a relatively small amount of matter, first as dust or mud, collects as a whirlwind with an eye (as would swirling water going down a drain). In theory, eventually such matter may agglutinate, evolve, and become defined as a planet with open poles and an equator perhaps to take its ordained place in our solar government. (See figure 35.2.)

35.2 — Growth of substantial particles out of dust, create a mini mass within a defined exospheric layer.

Speculatively, as the physical element of hydrogen is a constituent common to the outer Van Allen belt, could there possibly be a "hard" outer shell of known and unknown elements enclosing our universe with a collection of alien interrelated galaxies. Such a "hyper-shell" may be separating our universe from yet greater speculative universes as well of the Heavens beyond conceptual speculation?

— —— —

36 — OUTER AND INNER PAUSES

The shape of the pauses may be described as multi-capped mushrooms, or as a series of superimposed umbrellas, or as a tree growing out of a polar opening, with branches thickly spreading web-like in between the etheric layers, and likewise with the roots spreading in between the etheric layers within the Earth's hollow. The two trees of pauses reach straight forth outward along the Earth's axis of rotation extended, and each spreads out, all around reaching into the voids between the etheric layers belting the Earth.

The pauses host substantial amounts of dust, debris, water and other energies, siphoning these in either direction along the Earth's axis, and likely without any serious consequences. However, the magnetic and electric properties along the Earth's axis of rotation, and within the margins defined by the etheric polar cones containing the bundle of the axes of rotation of the etheric spheres, are not yet fully understood, other than they present a danger to aircraft coming to close.

(See Figure 36.1. An invisible tree comes though the north oculus, and another tree through the south oculus. The dash lines represent the two opposing cones delineating the trunks of the two trees, forming the bundle of etheric axes.)

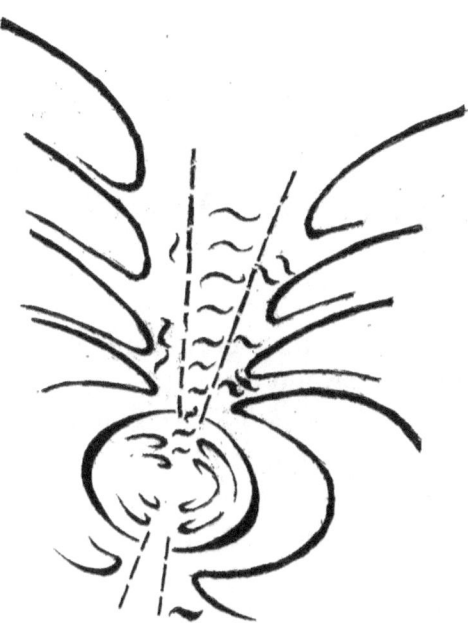

Figure 36.1 — Pauses, or space between atmospheric and other gaseous layers, combine into tree form above and below Earth's surface.

The pauses are characterized for their immunity against mingling and intermixing with the etheric spheric layers, and for holding substantial amounts of dust, debris, water, and so forth idly, of a different nature than of each etheric sphere, or charged differently to repel. The tree of pauses –branches and roots– of each planet, spreading out may be likened as well to a celestial spider web catching whatever it must.

Or, perhaps, as there are the invisible tellurian and electromagnetic currents and waves, or as nerve endings through animal red flesh, or as tree roots seeking and invading sewer lines to tap into available sustenance, and through tunneled pathways under the Earth's surface, there also may be invisible celestial bridge-ways and invisible celestial conduits high up above the Earth. Such may provide for more stable and safer traveling, than the contemporary piloting through a near unknown space filled with the invisible unexpected.

— ———— —

37 — PLANETS AND STARS

Rays, currents, ionization, magnetism, and assortments of cosmic forces, are some of the several distinct energy forms, the character of each a separate study. Simply speaking, ions traveling in the opposite direction of a current emit radiation as a perpendicular "third generation" resultant, and there are more, fourth, fifth, etcetera generation complexities, yielding either dangerous or beneficent results.

The stars are possibly no hotter and no brighter than the planets, and capable of safeguarding their perpetuity of existence. If they were hot masses, they would long ago have dissipated out of existence and thrown their galaxial environment off balance. Or, any dissipating heat should have seriously altered their arrangement and nature in general in all the cosmos.

Stars could well be planets that rotate in the opposite direction. It is known that all planets rotate in the same direction in reference to some supra-galaxial coordinates, and that all their axes of rotation are marginally askew to each other and bunched like flowers in a vase. If stars appear to turn in the same direction as planets, it could be because of the typical reverse flow of ions, or some celestial stroboscopic affect. Rays from planets encountering rays from the stars visibly and seemingly produce glare and a fiery reflection. Therefore, to Earthly eyes, the clash between the two oppositely ionized rays produces the aura of a fiery ball encapsulating each star. But, standing on the "planet" Sun, the Earth, and the rest of the Solar planets might also appear with a fiery aura, like a sun. Or, the fiery reflection may be from an invisible celestial honey comb-like division.

It is known that in going through space closer to the Sun, the solar rays are not as warm upon contact with an Earthly airship or spaceship, while away from the Earth. Spacecraft in outer space do not seem to suffer from the Sun's heat, but they do suffer during their reentering the Earth's "clogging" exospheric and atmospheric layers. On a sunny day outdoors, the heat is felt due to the ionized atmosphere.

The measured mass of the Earth or of the Sun, or of any other heavenly bodies, may not be read correctly through taking account of their gravities because these may be read and interpolated differently from different positions in space.

The opposing rotations between planets and stars function as giant gears, meshing and turning oppositely, which with the attracting and repelling forces engulfing each of the galaxies, keeps each of the Heavenly bodies in a strict and well-defined relationship — it is part of the fabric ordained by a celestial weaver into a celestial government.

The distances between planets and stars, the dimensions of their atmospheric domains and orbital areas, the rates of their velocities and speeds of rotation, and with their many particulars, have been known to Man from some primordial Golden Age. The Pythagoreans, having learned these from the Egyptian, Babylonian, and Persian priesthoods, revealed these matters to Western civilization.

The mathematic feasibility to working out these ancient lemmata are worked out neither on the base ten nor the base six numerical systems, but upon a forgotten system based on the "golden procession" (or Fibonacci system), a progressive sequence based on the ratios 0.618,033 and 1.618,033. Each being divided into unity produces the other. Modern studies prove that the ancient peoples of Mexico and Peru showed an equal proficiency to such celestial studies. (See: *Secret Code of Pythagoras — and the Decipherment of His Teaching*, by Hippokrates Dakoglou, in Greek, 1988; Athens, Greece.)

— ——— —

PART TWO

MAN'S WORLD

EARTH'S INTERIOR

*Protective soft bubble encasement of Earth, allows filtered
celestial nourishment and screens out toxins from outer space.*

38 — FLOATING LAKE OF FIRE

Upon entering the Earth's hollow from either pole, from out of the
cold and ice, and other conditions obstructing or permitting, with
a perusing glance at first one is confronted with an opposite scene.
At the first glance far into interior there appears in the interior sky
a lake of fire floating over a temperate surface of green and blue
with occasional spots of yellow, not too different than the familiarly
populated outer surface. And looking down where one stands, it is
a safe and inviting prairie nature preserve.

Advancing further, beyond the entrance and away from the
outside solar light, the lake of fire is discerned to consist of a few
distinct forms: a suspended little inner sun, a solidly fixed ellipsoid-
shaped white cloud straddling the axis of rotation of the Earth, some
lightly tinted lose clouds of irregular shape floating around and about,
some very faint high and low atmospheric demarcations, and some
birds joyously flitting about and reflecting their beautiful colors.

The solar and other rays of light, heat, magnetism, vaporized
matter, dust, protons, electrons etcetera, penetrate the Earth's
exospheric, atmospheric and terrestrial shells, and into the interior.
These gaseous shells forming a sequence of compounded lenses also
affect the environmental layers of the Earth's hollow. Some rays

ultimately converge toward an undulating focal point. The focal point is not at a fixed location, because of the ongoing oscillating and refracting irregularity due to the sequence of terrestrial plates and inner gaseous curvatures. Thus, by avoiding a sharp focal point, the danger of a possible nuclear blast and, or implosion is avoided.

Much of what appears at first sight as a lake of fire, along with the inner little sun, is an internally suspended atmospheric coagulation of matter from above and below the Earth's surface, resulting from irregular heating and cooling waves, luminous and flammable gasses, and dust and other essential and waste matter. All is composed into a beautiful scene of playful color and exciting motion in the metabolic process thereof.

The focusing, the hot spot, generally falls short of the Earth's center, short of the axis of rotation, more so when the Earth is more oblate.

As the Earth rotates, the inner little sun traces a spiraling high and low path inside the Earth's hollow (as turning on a lathe left to right and right to left), according to the inner seasons. During the Earth's mid winter season, the fiery cloud's circular orbit inside the hollow of the Earth theoretically comes closer to the pole of the winter hemisphere (of our era), and as the seasons yearly reverse, its northward path shifts southward, crosses the equatorial plane and comes closer to the other pole. (See figure 38.1, seasonal path of the little sun.)

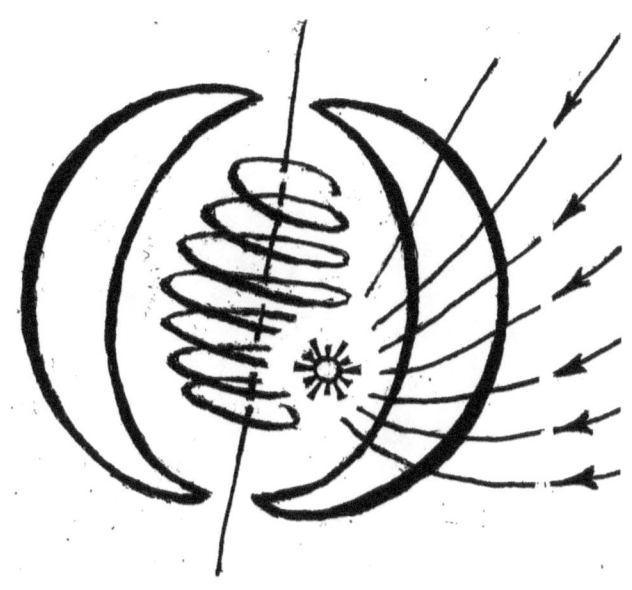

38.1 — Hot spot forms little Sun in Earth's hollow, spirals back and forth between poles, with path reversing seasonally in the hollow.

Perhaps the aurora borealis relates to the motions of the hypothetical lake of fire. Perhaps contributory to the cold of winter on the Earth's surface is the flow of ions from that half of the earthly shell to satisfy the internal centralizing hot spot, per the Kirchhoff theory.

If the Earth were a perfect sphere, the hot spot, inner little sun, might fall distinctly on the Earth's center, on the axis of rotation; it would remain fixed at the midpoint between the poles and perhaps be of milder intensity and with fewer clouds. Then how different and stagnant would the internal environment of the Earth be? How different would be the environmental experience upon the Earth's surface? And would the Earth's outer atmospheric and exospheric layers be different as well? Excessive concentrations of heat and radiation could be destructive or partially beneficial. Most likely, if the Earth were perfectly spheric, life inside would be next to impossible, save for lower forms of life.

Could some of the Mythologies concerning the underworld have been turned to negativism for the want of understanding and for security?

— ———— —

39 — PATH TO HADES OR PARADISE

There were times in the Earth's past when there were no glaciers and no ice caps. The warmer temperature at the poles and the slippage of the layers of the Earth's shell, had exposed one or both of the polar oculi.

According to Hellenic Mythology, Hades is inside the Earth, perhaps in the hollow of the Earth. In the Hades of the Hellenes, there is some form of life, but with gloom, rather a state of limbo, not totally dead, as they could respond when addressed. The etymology of the word Hades is "where nothing can be seen" – whether physically or allegorically. It does not imply hot or cold in Hades, but an acceptable temperature in dank dullness.

In the Hades of Dante, the "inferno", there is fire, punishment, torment and all the worst possible things just short of a neutralizing and vaporizing death.

"Per Mythology, if one goes to the ends of the Earth and falls off" —that is, falls into a polar oculus— "it is impossible to come out". But is that true? Could this have anything to do with the shape of the ocular lips at that time? Some religions tell of the weighted souls allegedly sinking through the terrestrial shell directly into Hades. Whereas those souls, which remain spiritually alive when physical death comes, it is said, levitate into the Heavens. Ancient records, as well as Mythology, will soon be offering scientific clues. Little understood in the immediate past, in Mythology there are the rich records in symbolic form of the events prior to the Ice Age. Certainly, interpretation is needed to calm the negativism of fear. Nevertheless, the world is still alive!

Perhaps it is not a path to Hades but a favorable path to Paradise at that time, a land equally capable of bliss, as on the Earth's outer surface. Or is Hades delimited to a certain geographic region in the

Earth's vast hollow? After all, on the outer surface, we do not hesitate to spend much time indoors for health and protection.

Many indications for life in the inner world are: a little sun, plenty of water and oil, earth to sustain flora and fauna, beaches, forests, therapeutic rays, and why not decent upright human life too? And, are there hunting, cultivation, fishing, agriculture, and a few intellectual attractions yet unknown on the outer surface of the outer world? The last is scary, for they may be more intelligent than we are and more sincere than our politicians. Innocently and imminently they may be exposing us and them with their treacherous New Age connections to our and their shame.

Why don't we send a peace-seeking mission down to them, as we do to small nations whenever there is room for our taking over, and exploit them, too? If our present big-power secretly oriented governments get them mad enough, they may come out and give us all a spanking to teach us a lesson on how to be good citizens and, really, how to vote.

— ——— —

40 — ALLEGED CORE IN THE EARTH

Each motion and each difference between the motions in our galaxy —the solar system, the earth-moon barycenter, the Earth's atmospheric layers and belts, the Earth's surface, the crust of its shell, the mass of its formed multi-layered shell, its interior surface, and inside atmospheric layers— is related to energies, whether active or potential and chemic or electric. Each of the layers, from exosphere to core, be they gaseous, liquid, or solid, functions as a dynamo within a dynamo, and in succession within dynamos, generating new energies.

The "pauses" between the terrestrial layers absorb most of any friction and serve as insulation, not too different from the "pauses" serving outwardly from the Earth's surface and above.

Within the hollow of the Earth, hypothetically, from the center outward, there are: a near perfect vacuum, a frozen hollow sphere (or cylinder) of hydrogen, the orbiting inner little sun, and the inner layers of atmospheres —perhaps with occasional clouds— extending, to the inner surface.

Pauses and atmospheric layers could hold the inner little sun in suspension inside the Earth. At the closer confines of the Earth's center, close to the center of the line of the axis of rotation, at the core of the earthly bubble, there may be an extreme rarity of atmosphere and much cold. The core, ellipsoid in shape with long axis straddling the Earth's axis of rotation, while lying hidden from the naturally foreshortened focus, is the least affected by all motions and energy. The exception is some vestiges of draining of some heat out from the core, per Kirchhoff's theory, super-freezing the core, while magnifying the heat and brilliance of the little sun. A near vacuum may be generated by the unencumbered centrifugal force, and by the cold, because of the drawing out of ions into the adjacent atmosphere.

The inner little sun is not in the center, just as the Sun may not be in the center of the Earth's ellipsoidal orbital path. The centroidal axis of rotation of the core, from collinearly on the axis of rotation of the Earth shifts in position, causing it to tilt and wobble irregularly, relative to the axis of rotation of the Earth's shell.

The most rarefied of all known elements, and the least affected by centrifugal force, is hydrogen. Speculatively, the ellipsoidal core of the Earth is of frozen hydrogen. More likely, the core of hydrogen may have a substantial hollow as well. The mass of the hydrogen core varies with all outer changes, as expected. The hydrogen is the result of its separation from the waters pouring in through the polar oculi. The core of hydrogen is cumulative but is kept in shaved control by the little sun traveling helically between the polar oculi, like being turned on a lathe. (See figure 40.1, with rays from a high summer Sun.)

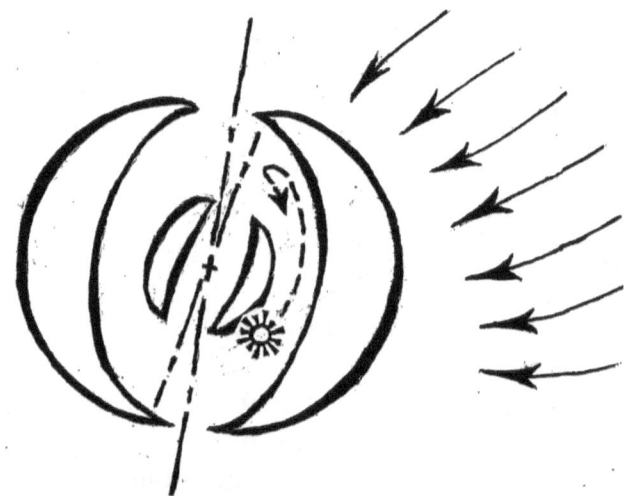

40.1 — Suspended frozen crystalberg inside the Earth keeps its shape by a free spiraling interior little sun.

The dislodged particles of hydrogen are thrust outward forming a cloud and then combine again with the waters in the interior. From the vicinity of the core a dark cloud of hydrogen with other matter appears, but which at times is open to light from either or both poles?

The core of frozen hydrogen may wobble off center from the Earth's centroid, just as the Sun is off center from the Earth's anomalistic elliptic orbit. In addition, the spiraling orbit of the little inner sun may be anomalistic. Perhaps the magnetic axis goes through the center of the little sun or the frozen gaseous core rather than relate to any other terrestrial phenomena.

— ———— —

41 — OUR DAILY PETROLEUM

Why do we continue discovering new petroleum and gas deposits as the older ones seem to dry up? How much petroleum and gas has been consumed since the discovery of these commodities? Were there that many overweight dinosaurs running around, to accommodate all our automobiles, etcetera?

If in the metabolism of the human body, the carbohydrates can be converted to fats and vice versa (fats into carbohydrates), as the dieticians tell us, then the Earth, too, can produce abundant amounts of oil (as oil is fat, too), daily and hourly. As the Earth's external evaporation and precipitation is in waters, possibly a form of internal evaporation and "precipitation" may be in oils. For nature's processing, the ocean currents bring huge amounts of garbage with fat and other organic matter, in a constant supply, into the Earth's interior through the polar openings, if not by nature's alternate passages. Here and there, from certain internal zones, the oils find or filter their way through the terrestrial shell's layers, being refined under great pressure in the process, to the Earth's outside surface. Petroleum may well be one of the very plentiful by-products of terrestrial metabolism for which there can never be a shortage. And the vital vegetable oils are drawn up further by the Sun for the nourishing of most of botanic life.

The wobbling and shifting of the floating core of helium in the Earth's center and the orbiting of the little inner sun, and in general all terrestrial undulations, serve with a pumping action to accelerate the rising of all kinds of oils toward the Earth's surface. As well, they serve to move the subterranean oil pools around, drying up some regions, and opening through in other regions, instead of replenishing certain known productive regions.

Major petroleum and gas veins under the oceans and lands lead to discovering these deposits of sundry commodities, as well the sources of springs. These may be made accessible, further developed, and exploited throughout the entire Earth for Man's comfort and enjoyment.

— ——— —

A question: If legal precedence is set up for monopolies to extract gas and oil out of "Mother Earth", then why cannot licenses be granted to monopolies to extract fresh drinking water from the same "Mother Earth"?

A most obvious fact for questioning: If Banking Vipers periodically go to war against each other, and if oil and gas monopolies as well go to war against each other, is it possible that most of the "Oilies" and most of the "Vipers" are out to destroy each other in a game of complete domination through, say, supply and the tricks of demand? Will the "Water Boys" play a role as the third belligerent party in the melee for the control of the coming oligarchic New World Order?

— ——— —

Cutting through plates and layers of the Earth's shell there are sinews, energy (cerulean) lines, mineral and aquacious supply line, oil and gas lines, and sewage (lymphatic) lines. Earth movements, whether natural or Man induced, as by bombs, subway construction, deep mining, and rising of tall voluminous buildings variably apply pressure or release to any of these lines, opening or constricting flow through them, not without effect on the life of the Earth.

— ——— —

42 — ANY LIFE IN THE HOLLOW?

Can it be proven whether any life could be sustained on the inside surface of the Earth? That is, life in the form of flora, fauna, bacteria, etcetera? Yes! And as there is life in the darkness of the grounds we walk upon and in the great depths of the oceans, of course there is some kind in some places deeper within. For if life-bearing waters counter-flow in through the polar oculi, there are unquestionably hydrogen and oxygen, as well as alternating light and dark for photosynthesis. More so there is life, as the Sun shines through the polar oculi, be it in its full glory, or dimmed, or clouded somewhat by the varying condition of the oculi through the different geological ages. Life, then, must be just as active to varying degrees within. Perhaps more light is generated from internal chemic and electric activities. Particularly, forms of life may be the same, or similar, and some yet to be discovered.

Even at times when the polar openings are shut off, the abundance of electromagnetic energy yields plenty of pleasant natural calming blue light in the interior.

— ———— —

There it is, the little sun, orbiting around an ellipsoidal "crystalberg" around the Earth's axis of rotation. The little sun shines joyously on beaches, meadows, mountaintops, back yards and most everywhere upon its inner domain.

"Seasonal" changes? There have to be. The "seasons" are related to the north oculus's leaning away and toward the Sun cyclically through perihelion and aphelion. The temperature and radiation within must vary with time and place, no differently than on the outer surface. Of course, there must be life within, so that the little sun therein may be enjoyed in splendor.

Can there be a daily sunrise and sunset within? Very possible – rather, yes! Then perhaps there is a little rainfall, but always there is pleasant moisture. The massive core of frozen helium floats in the center of the hollowness. Perhaps it is mixed with other life-sustaining gases as well, mindful of matter in the frozen polar ice caps. This massive frozen core, as a solid and permanent cloud, cyclically hides the shell's inner surface from the little sun when on the opposite side. As the little sun travels around in its hollow domain, it causes inner eclipses, alternating warmth and coolness, and the comfort of sunrise and the reprieve of sunset. The little sun rotates in the inner atmospheric margin between the Earth's inner shell and frozen core, while it winds seasonally north and south toward the one pole, and then the other pole. The timing of the seasons may well be affected with the rhythm of perihelion and aphelion of the outer world, and as well to the totality of the holistic astral influences with the heliacal rise of the Star Seirios, to give variety to an agricultural life.

With what green and greened lands and what blue waters? Snowfall and hot seasons, spring and fall? Trees, creeks, roses, melons? If all these are givens, why can't there be any interesting life within? Hunting, fishing, swimming, picnics, entertainment, popcorn, skating parks, gazeboes with musical bands in town squares, Zeppelin rides, Ferris wheels and other delightful contraptions of entertainment, too? Why not giraffe races, brontosauri in tugs-of-war, and giant turtle wrestling? Now imagine, there are staged entertainment features, of comical talking birds, whistling fish, laughing mice, and donkeys in arguments over sports and economics, and dancing monkeys collecting pennies for their drum-beating masters? And as for radios and television, rarely it is announced that they will go on for a certain vital program, for their level of education, social decorum, and established local talent for the lively town square gatherings around the gazebo.

Any human life? That is unknown. Legends lead some to believe that human life may be sustainable within, or that actually there is such. Others believe of grave moral negativity therein. Others claim that there is an ongoing war therein between the good and bad, like on the Earth's surface, ongoing since Mythologic times until today.

To complete the settings, should there not be an inner main zoo with animals from both the inner and outer surfaces, safe in their cages, from which the animals can get to watch in review all the curious human people going by in parades before them? Dinosaurs give rides to whole families on their backs through thrilling swamps with acrobatic crocodiles making their passes would be most educational.

Schools, factories, research, interplanetary exploring expeditions to the innards of other planets and stars, as well to arrange interplanetary marriages, provided they find any people-kind of life there in the insides.

— ——— —

Alas, there may be evil and justice therein too – hopefully less evil and more tranquility. Why not cops and robbers sensationalism too? An interesting bit of social history is rumored, cleared and so published:

> *Associated Inner News Flash:*
> *The Inner Detective Service for Social Morality captures galaxial fame for discovering the passage to the Mongolian Dessert from the Inner World. Legend had it that Alexander the Great sealed up this passage for the protection of the Outer World, per instructions he received in Ammon, Egypt. Subsequently, it had been reopened for trafficking vials of misery, which were poured upon humanity of both the Outer and Inner Worlds.*
> *The Inner Detective Service discovers their lair, exposing and dismantling the entirety of the secret Inner Headquarters with long files of evil doings. These files*

were thought to be secure in the remnant off-limits Nephilim Quarters of the Inner World.

These files belong to the outer Internationalist Banking Vipers, including such associated cabbals as the Federal Reserve System, the Royal Reserve System, the Red Commie Reserve System, Wall Street Reserve System, Fascist Reserve System, Knights Templar Reserve System, Free Masonic Reserve System, Jesuit Reserve System, etcetera, all syndicated as the "Sell Short Reserve System" for the more efficient fleecing of all the peoples.

Evidence shows, these Vipers are the major cause of racial, religious, social, economic, judicial and educational upheavals, and indeed the reeking of poverty, misery, and bloodshed on the Outer Surface for the last two and a half millennia.

After a few minutes of due process, the Ministry of Inner Security locked them and their notorious Boards of Assassins up for life in zoo cages distributed across the entrance square of each inner regional zoo. They are featured as the most dangerous animals in our galaxy.

Their grand master, Satan, immune to radiation and phasing into invisibility at will, escapes. The galaxy police are put on notice to trace his hoof-prints and tail-sweepings wherever he tries to set up new usurious Viperages, to perpetuate the Negative Monetary System, which talks to Man, rather than serve, as well now preparing to re-conquer the Inner World.

However, there is on the Outer Surface

a movement of pious misdirected innocence to save their endangered Vipers with their awesome Viperages from extinction: "Let us save our holy Vipers! For humanitarian and religious reasons! We need to borrow more of their most precious paper money to restore our sweet happiness! They are so willing to loan out to us all the money we need to enjoy life!"

Their haughtily loaned-forth magic bloodsucking printed devices, of numbers, subject to flames, washouts, or rat food and called "money", is still adored as the practical object of worshipful success. Winners of such are crowned with "arrival" to the exclusive rooting out of real universal happiness and the eternal salvation of their everlasting precious human hides.

The conspiracy of the Banking Viperages was discovered as they began pushing into the Inner World their Satanic Negative Debt Reserve paper money (which demands the yielding of interest out of nonexistent assets, akin to "selling short").

After having retaken and restoring to the Titans the rule of all the Outer World's civilized nations, now as obedient Debt Prefectures (no longer as free Nations and free States), the Vipers imagine retaking the Inner World would be their next and easier well-deserved prize to serve up to their viper-gods in exchange for rewards.

Well, that, too, is another subject. Back to our theme.

Could the changing and rearranging of the terrestrial plates expanding and closing the polar entrances have arranged for the eviction of Man from Paradise, eventually for the demised void to be converted into an incarcerating Hell, worse than before? The undersigned pleads ignorance. Dr. Rohrschach might lead us to some answers.

– ––– –

On the outer surface, when standing on flat ground, we sense that we are on slightly raised ground, as distances disappear downward left and right beyond the horizon. When on the inner surface, on relatively flat ground, imagine, we sense we are in a sweetly swooping grand valley, as distances fade upward and away to a haze, perhaps to beyond visibility. Viewing distant regions within, when nighttime has arrived, the slowly encroaching nighttime can be experienced as the inner little sun is disappearing upward beyond a covering haze, or cloud, or behind the emerging "crystalberg" of helium.

Little golden-orange lights of the evening from the more distant habitats exude happiness as people enjoy life by singing of life's pleasantries just before their bedtime. The encroaching day-time can be viewed as the little sun comes out of hiding from behind the suspended central "crystalberg," slowly warming the faces of its viewers.

As flora and fauna differ on the outer surface between the continents and latitudes, so such must differ on the inner meadows, plateaus and islands. How different? All within are subjected to less cosmic radiation, for the Earth's shell serves as an additional protective screener, but for the internal radiation, as all the beneficial elements are prismatically refracted from the workings of the little sun. Perhaps intermingling and interbreeding of flora, fauna, and of human people with the outer world may all work out advantageously.

– ––– –

43 — GRAVITY CENTERED WHERE?

When entering the Earth's interior through a polar or any opening, do we experience a free fall into the center to get burnt or frozen? No, neither! We merely walk around the edge, onto the inside of the Earth's shell. We look overhead to see the little sun or wait for it to "rise" from the other side of the cold mist, or the crystalberg of helium. Should we fly in through the center of the opening, we will gently be forced laterally to land safely.

Gravity, on the outer surface, may be explained as a centripetal force, but on the inside surface, it may be explained oppositely, as a centrifugal force to maintain attraction toward the inside surface of the terrestrial shell and to keep people, trees, bicycles and animals steady on their feet in the inside world. Looking up in the outer world is diffusive; looking up in the inner world appears concentrative. In both cases, it is assumed that the effective difference in both forces is related to the centroidal or median gravitational layer, a median sphere, webbed between the outer and inner surfaces of the entire terrestrial shell. (See figure 43.1, exaggerated, ice cap shaded, generally theoretic).

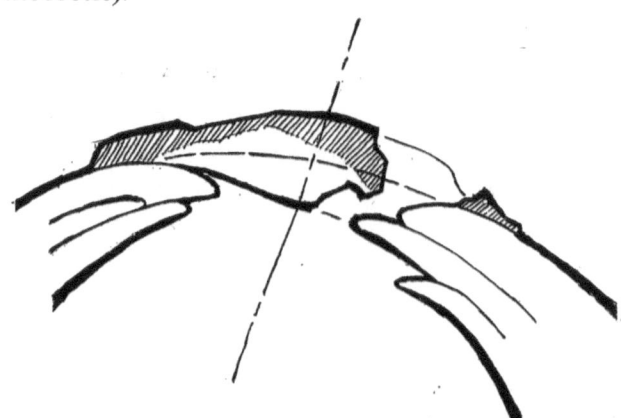

43.1 — Imaginary section through Earth's polar opening with ice cap as it narrows the passage to the interior.

As for Man's mastering the Earth, one must be reminded that even on the outer surface, human life is sustained with technological advancements at geographic regions formerly deemed uninhabitable, but today marvelous shelters are set up within the Arctic Circle; there are air-conditioned automobiles in the equatorial deserts, and life in submarines is extended for months underwater. Man's satellites hover in space for months before returning for fresh air, water and victuals. The terrestrial outer surface is almost finally being "conquered" by civilization.

Geographic examples: Life in northern North America was not able to flourish into an advanced civilization until the advent of the age of technology. It was a struggle to survive the extremes and suddenness of climatic changes before the age of technology, such as to suffer from undo exposure from the dropping of forty degrees temperature within the hour, turning from rain to sleet and ice. The drying of the marshes on the steppes by the Caspian Sea allowed continual nomadic invasions westward and all-season caravans. The speeding up of the Earth's rotation will cause more dangerous turbulence in the seas separating Antarctica from the Cape of Good Hope and the Straits of Magellan. So what, today we have airplanes that keep us in a comfort zone over these obstacles and through very cold high altitudes.

After Man has learned as much about the interior of the Earth as he already knows about the galaxies, he may overcome the obstacles to transit through the polar oculi, or other passages, and become accustomed to life within the Earth, exploiting the possibilities for food and wealth in square deals of exchange with our own "Inner" friends as well as new physical elements to make the outer scientists curiously happy. And, perchance, if need be, to chase the devil out, to find another abode for himself, perhaps in some "black hole", his new hell, off in a far-off corner of outer space.

Nonetheless, there have been some rumors, being repeated through the centuries and in our more recent times, of explorations and accidental discoveries of an opening within the Arctic Circle to the Earth's interior. It is said that having traversed inhospitably cold and desolate areas, a warm valley is discovered with flora and fauna. Properly, such should belong under the Earth's more temperate zones.

Warm out-flowing currents of water and turbulences are noticed, per unbelievable stories around the heroic and gentle Amundsen of Norway. Such valley supposedly leads to the Earth's interior, from which light is seen with an extension of possible life. (See figure 43.2, exaggerated; compare with 43.1.)

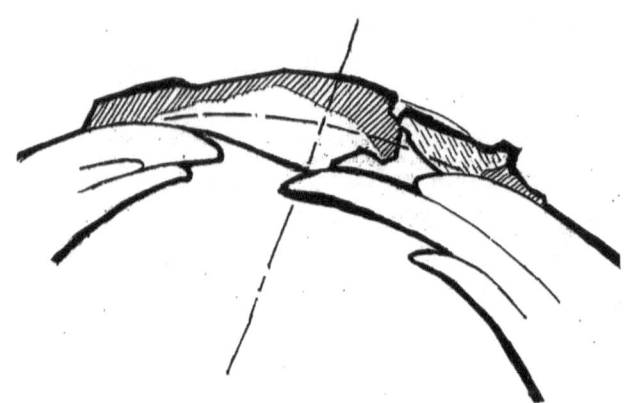

43.2 — Imaginary section through polar opening of Earth, which blocks the ice cap and a shifted plate seals the passage to interior.

Other reports give that undigested flora and fauna are being discovered in the bellies of long-ago animals suddenly frozen in place, whose habitat is the presently frozen Arctic Circle, due to a sudden tilt of the Earth, or due to whatever we are allowed to imagine.

It is possible, in theory, that several valleys and entrances to the interior of the Earth periodically open and close, according to the shifting of the suspended massive polar ice caps and the shifting of plates composed of the Earth's shell, or as well anywhere else on Earth.

Is gravity a form of centripetal force?

— ——— —

44 — WHAT IS LIVING LIFE?

Man differs from fishies, birdies, animals, and bugs, in spite of the short lifespan of individual human beings, in that Man, that is, healthy Man, instinctively is capable of and wills to perpetuate his cumulative culture with his skill, rational memory, and last testaments, for the future generations, with all its accumulated seriousness, bounds of humor, registry of sins, merciful acts, repentances, failures, and indeed with a summation of worthy accomplishments.

To speak with practicality, one observes that those sophisticated professional "doubters" of the richly endowed hyper-complex environment, given to us from Above on our behalf, are speculators on quasi-truths far beyond the probability of veracity according to human social tradition and reasoning.

Aside from the origin of matter already alluded to, what is conscionable or noetic life? How has that begun? Atheists try to explain intelligence by their expanding on gasses and convoluting on sparks. But such is inconclusive and far more nebulously speculative, than the theology they try to belittle.

Their most sophisticated, egocentric way of trying to begin to explain living life, or a living soul, is by their uncalled for "spark", or unexpected "bang", sans derivation, sans provocation, sans intelligent Giver of the direct order to spark! and bang!— too simple, on a level for monkeys to grasp. And recently, again the Atheists make their fanfare, this time with their "jello theory", that the universe simply and suddenly congeals quite uniformly throughout, again, out of absolute nothingness, no causal order or causal action, into a twittering existence, twittering kaleidoscopically into forms to be forgotten the very next nanosecond, all again no differently than previous Atheist theories, all being without "cause, without purpose, and without means". That is, without an author, without an objective, without any means of potentiality. The ancient Hellenic philosophers conclude that nothing begets or gives birth to nothing.

— ——— —

A basic "Chapter One" of the origin of all that is routinely is missing from books on the sciences and history; for the universalized Solvey sciences find it difficult to begin with axiomata. Of those alleged sparks indicating a new existence, as seeds budding, cells duplicating and splitting, and sperms germinating aquatic, reptilian, animal and human life, rational religious folk ask: these are sparked off by whom, or what, if not from the Creator's Mind, for His Purpose and by means of His Energy (that would be the beginningless and eternal Holy Trinity, or collectively The Being, that is, by The Truly Wholesome Being), then by whom?

Formerly, among some pagans, the "Great Spirit" (American Indians), or "das Geist" (ancient Teutons, in neuter gender), etcetera, must be the prophesized Holy Spirit (or the Spirit of God); that is, perhaps after their having forgotten the prime cause and the objective as irrelevant, due to a primordial cultural convulsion coupled to the varying values within their tradition, in respect of the yet unknown God. It is one of the ways that the Creator enters His creation. But no way per the Atheists's religion, disorganized as it is, based on thousands of tons of speculations, trying to fight the Creator, can cause these "accidents" to "begin existing", not through any of their alleged "*deadwood material*" preexistence. Further, if in physics, the indestructibility of energy is true, conversely, how can a spark come into its beinghood out of what nonexistence, to bring forth formulated existing things? And, "*dead*", by what cause having died? Any kind of "*material*", and by what category of its preexistence?

Before trying to explain the origin of being, there is an axiomatic premise to be resolved: Does material precede Spirit? Or does Spiritual Being precede material being?

This leads to a second question: If spirit is simply another form of natural energy (rather than Divine Energy), which, too, is a product of this world, how can it break the "scientifically" established rules and perpetuate on its own non-conscionable cognizance a generated being of greater complexity? That is, how can it violate a cycle of ice, water, steam, and ashes (which still are, though metamorphosed, in the created world of being), and yet leave no historic trace of any conscionable evolution? How dare they violate their very own laws of physics and of nature to come up with an anti-axiom, to promote what? Although a challenging question, the answer, if possible, would lead to beyond the objectives of this work.

— ——— —

Being an Atheist sadly is denying one's conscience on his personal existence. More sadly, when such Atheist expires to be consumed by nature, can he kill his indestructible conscience, to stop its existence, so that it may not haunt him in his metamorphosed state?

In Byzantine academic thinking, there is the division between —but in a working relationship— of philosophy and theology. Philosophy axiomatically begins upon the act of Creation, by which the element of time begins to exist, and comes to end for whatever ceases to exist, if not to end upon the end of universal existence. (For instance, old clothes cease to be of use and are chucked as worthless; their time as clothes thus having come to end.) Theology covers from before the Creation, or preexistence, goes through the length of time consciousness, and continues beyond the speculated end of time, metaphysics. There being no conflict between Byzantine philosophy and theology, the domain of each emerges clearly defined. However, the values or axiomata of each, when found to be in concord, is taken as proof of the validity of each, respectively toward perfection. Emperor Justinian, aware of such relationship, establishes a legal system (known as "Roman Law"), cautiously as a third entity to the standing bipartite concord.

— ——— —

EARTH'S RELATIVITY

Rhythms of Earth's flexibility maintain joint celestial and terrestrial integrity with rhythms of the Cosmos.

45 — FLEXILE EARTH AND CHRONOLOGY

The endless variation of the Earth's motions leaves an enigma for the creation of more exact measurements and for the formation of an ideal cyclical calendar, seemingly an impossible problem to solve because of the many and almost arrhythmically and mutually influencing cycles, some of which never fit clearly into nodes. Some fit but with rhythmic periodic intercalary adjustments, as by one day over long periods of time, while some distinctly fit into obvious nodes. Groups of cycles relating repetitively to greater cycles are the nineteen Olympiads to form four Metonic cycles, or one Callipic cycle, of 1,461 weeks, etcetera. In each of such groupings, there is an odd number of days fitly and rhythmically to accommodate the "Seiriac leap year", which is a regular occurrence each four years. (Seiriac pertains to the star Seirios. See *Julian Calendar Valid*, by this writer.)

The long-term problems include not only the changing number of days in a month, year, etcetera, but also the changing of the duration of the very hour, month, or year. If there is any uniform expansion and shrinkage in our galaxy, in reference to extra-galaxial parameters, then even the most basic unit of measure must be variable in length and weight, along with all the other units related thereto an undulating, but then, do they lose their nodal cyclings? Or perhaps, there is a jumping in the nodes, as in the shifting of gears for the differential speeds in manmade engines.

An elephant doubles its size, that is, in three dimensions. His volume and weight are sixteen times greater; but his footprints come out to be four times their original area. Can the area of his feet sustain his new overpowering weight? Likewise corporations, nations and all that is under earthly, or centralized control, reach a certain limit and begin collapsing bottom upward (like Daniel's giant, shattering first in its clay feet and progressively shattering upward to its golden head). There are the natural, the human and the metaphysical scales for Man's contentions.

Consider a simple truth: a segment of a circle of thirty degrees at a given radius draws an arc of a certain length, and another segment, also of thirty degrees, but with a radius twice as long, has a greater length of arc. Accordingly, the days closer to aphelion, although through each thirty degrees traveling along a longer arc, require a longer time, a fractional day more, or an extra full twenty-four-hour day, to keep at the thirty degrees. (See in Chapter 14.)

Philosophically and hypothetically, the Earth's changing configuration and motions may be studied in two ways:

The Earth, as with all planets and stars, is in a constant state of arbitrary compensating reaction to the multitude of sundry influences from propelled and impelled accidents, reactions, and random happenings with reverberations. All of such assure perpetuation, even in the absence of "absolute" government, as defined by Man. In this first hypothesis, any absolutely rigid and inflexibly solid planet, if ever subjected to influences beyond any capability of compensating reaction, would explode or implode into dust. Or if not, it would crash with another heavenly body, or disintegrate into ashes on approaching a heat zone.

Or, per a second hypothesis, the Earth works with all forces rather predictably, instantaneously and in harmony, as a part of nature's universal government, including the explanation of inertia and friction of least destruction. Such hypothesis would sit better with any hollow Earth. In this second hypothesis, there is some invisible but highly sensitive bond with all the heavenly bodies, being endowed with sufficient cushioning between each. A hollow planet can function in a universal compensating government dexterously, without encumbrance, and with the least of inertia, as though a balloon floating in the air, bouncing about most delicately with some transitory resistance. Or else, a hilariously amusing or a most frightful noisy big bang of termination would be expected.

The changing number of minutes and seconds of time, assuming they are variable units, in the length of each year is neither on a steady rise nor decline, neither acceleration nor deceleration. But these units of time manifest in alternating surges, covering millennia of years, perhaps impossible yet in this age to detect and calculate, and to establish whether a waning or waxing of the undecipherable long-term nutations. Minor surges over a long era may be misconstrued as for major changes within a single century. Except for acknowledging that there is a galaxial and heavenly government, it is impossible to establish a formula for an eternally absolute calendar in the physical sense, but as for a relative type calendar of standardized fixed measures coming into nodal agreement, or harmonic agreement, it is possible.

One may hear from scholastic oriented scholars, that all calendars are man-made; therefore, it makes no difference on how a standard calendar can be set and standardized. Wrong! Nature's God gives all calendric concepts, timings, cyclings, rhythms and divisions; Man merely employs his phonetic alphabet (or hieroglyphics) and abstract numbers to define what nature is showing him.

— ––––– —

46 — PLANET UNDERSTOOD

The word "planet" is from the Hellenic "planetes" and means "wanderer". It is derived from the verb "planao," meaning "to wander about, to stray." The word "aeroplane," or "airplane," is derived from the same root. The Hellenic noun "planetes" is in the masculine gender, implying an active, predetermined, and generative role in the planet's inherent mobile traits, rather than its being a mute object under a controlled suspension to be guided and drawn in any given direction.

The ancients knew that a planet is not simply a missile being directed through space. It is an organic and energetic body, which floats and conformingly guides itself by its innate ability to parry recklessness and chaos. A "planetes" navigates through a flowing medium, while having some active determination through the fecundity of its physics, and confines itself to a defined area of operation, according to a greater celestial mathematic government. A ship lost at sea does not imply that it has sunk or that it can never again find its direction.

The word "plane" is feminine, another derivation of "planao". In Hellenic, the feminine noun plane denotes passivity, as to be led forth or led astray. In English, plane has two meanings: one, a two dimensional geometric receptor; and two, safely hosting a wandering object within the confines of two directions. Philosophically, two dimensions may indicate any direction upon its plane, whereas a single dimension is simple direction, a pointing, and defines a location on the plane, as of a planetary axis of rotation delineating a circumference, which in turn defines a limited plane.

Before the Dark Ages, the ancients knew that a planet is a heavenly body, "running around in predetermined circles," under a government within a plane, the position of which is given relative to the Heavens. The ancients knew that each planet or star is not an object randomly lost in space, or glued on the ceiling of space, but all come into a "cosmodomic" plan, or a "universe planning" (akin to "poleodomics", or "city planning").

Is a plane merely an abstract concept upon which a direction, or trajectory, or path of travel can be plotted within the limits of two directions? The lunar plane being askew with an ever-changing angle relative to the orbital plane, and the twirling of the barycentric arm everywhichway between the Earth and Moon, would say, no.

Then, is the Earth's orbital plane an infinitely extending celestial field with a determined thickness, say, as a sheet of ice having a certain thickness and function, as to bear the given physical activity of a given object on its travel? Or, more specifically is it a sheet of energies bearing a planet or group of planets with their activities? Should a planet's momentum weaken bringing a planet to a standstill, would it be lost at space, or would it be vulnerable to be picked up by another heavenly body's magnetic force, or for it to entrap another body to its sphere, and together continue as a duet to orbit upon on another plane? The lost in space satellites may lead to a clue.

Per astronomic measurements of moving galaxies, which are light-years away, the triangulation of their arcs in length and speeds reveal calculations faster than light. All in creation is relative —in the undersigned's opinion— to a higher orchestrator.

— ———— —

47 — RELATED SCIENTIFIC CONCEPTS

It is fashionable to discuss antimatter, antigravity, and anti–everything else with more speculative qualifications for new terms.

Just as antimatter is not a lack of matter, but beyond that, of an opposite kind of substance, antigravity is more than a lack of gravity, not a spatial rest but an opposing or lifting force. And so darkness is not an absence of light but a substance opposite of light, a substance projecting darkness and swallowing light (Apostolos Makrakes, philosopher, theologian). This should not be strange, for in the range of wavelengths, there are the visible spectrum and the higher and lower invisible ranges of wavelengths, sensible to our other senses. This may be illustrated by a wheel with spokes, which appears to be turning backwards at certain speeds due to the stroboscopic affect. And, among many cultures, nature is known to react negatively to Man's will, when he is out of rhythm, when unsavory.

Animal, air, and aquatic kingdoms have developed senses beyond human capacity. Humans as of yet do not know the body languages of all such living things to interpret their communicating with others of their own species, with Man, and with other species. Do humans know much about their own body language? But strangely and to our humility, other living things, flora and fauna, can read human body language, and perhaps human thought.

Black holes, sunspots, etcetera, may be opposing counterparts to such phenomena, as is the aurora borealis, or the hot spots of glare formed by concentrations of light rays. Perhaps planets and stars likewise are of opposing phenomena, perhaps a reversal of positive and negative. Velikovsky suggests the difference between stars and planets is merely in that they rotate in the opposite direction and that they repel each other, creating the blinding phenomenon. But perhaps stars as well may be of a different unknown consistency, neither planetary nor gaseous. If such environmental fields could be bypassed, the realm of the stars may be physically entered, when technologically permissible.

"Black holes" in space are, supposedly, locations in celestial space, which entrap and swallow all matter in their way. This means, theoretically, that a cubic inch of black hole must weigh millions of tons or more. How, then, do they maintain their suspension? Do they convert into a suspended "black" energy to stabilize a celestial balancing act? Is there a trinity composed of planets, stars, and black holes?

As the stars project light containing energy and vitamins upon the planets, likewise black holes perhaps project upon the planets a darkness containing mysterious powers (little understood, but not evil) of growth, recycling, etcetera. Thus a Trinitarian image of the Creator is projected into the physical world, just as Man is of a Trinitarian make, of body, soul, and spirit.

On Earth, it is observed, there occurs an inverse black lightning. It precedes the visible noisy lightning bolt, zigzagging upward to strike the cloud first. It reaches the sky from Earth, and then on the same path, the bright lightning striking back down.

Air travel evolves in the following stages: A glider floats through the atmospheric medium on the air currents by momentum and gravity. An airplane moves forward as its propeller drills forward into the atmospheric medium, pulling the vehicle forward. In jet propulsion, the vehicle is pushed or hammered forward into the medium, creating a much greater pressure behind it then the resisting pressure in front. The next stage is in generating a leading vacuum in the direction of the path, by rays, vibrations, or instant compaction of atmospheric energies into a reduced volume, into which the vehicle is drawn (sucked) forward.

"For every action there is an equal and opposite reaction" may be true in an isolated and static environment. In an intense environment of other motions and energies, the reaction (rebound) may not follow in the same length of time, but may be delayed even by centuries. Hence, hard terrestrial matter may actually be proven most resilient ultimately, but untimely relative to our detecting senses.

In today's computer sciences, especially in using the binary system, "rejection" and "abstention" are confused, both being assigned the value "zero," while "acceptance" is represented as "one." The astrophysical sciences especially are handicapped until the scientific mind can see more clearly the use of a trinary computer system, with a "plus one," "zero," and "minus one." Would the Zoroastrian-oriented academias of today immediately offer resistance to exploring for a trinary system?

Seeing how nature works in the heavenly firmament and on our Earth, many axiomatic lessons may be derived for advanced technology. Force, pressure, friction, etcetera, could be replaced by materials assuming changes in size, shape, and texture "strategically" through their own inherent potential, upon being addressed by a combination of critical fields of sundry energies.

The flexibility of all fauna and flora relates to the flexibility, or to the reactive flexibility, of the planetary and astral motions and functions, which in turn relate to the conscientious morality of the race of Man. And How paradoxic is it, that Man today fears nature more than he does God, while in reality Man controls nature, optimally when in proper order with himself, with God mercifully (per the pre-Christian Socrates, believe it or not) and with good will overseeing.

According to Byzantine philosophy (N. A. Matsoukas), nothing under either of the categories, non-created or created, is evil. However, of all that is, the perversion or misuse of anything or idea becomes evil. That is, all evil is secondary to Creation, an unfortunate derivative from what is good. It is an optimistic and hopeful system. A secondary class of ideas and beings may be corrected and restored to primary goodness. Created matter and energies are subject to the element of time, a beginning, a duration, and an extinction.

What about natural violence and catastrophes? Some involve people, which are in the image of God. Some such happenings from before the creation of Man are not worry-worthy. Man, which potentially is in the likeness of God, is endowed with the freedom of conscience —not to be harnessed by authorities, neither by dictators nor secret oaths— to understand and conquer the universe, to protect his fellow Man and enjoy all that is for the oneness of mankind.

— ――― —

48 — EARTH AND EXTRA SOLAR SPACE

Motions of the Earth, described as relative to the Sun, are: Earth's travel on its orbital plane; Earth's spinning about its axis; spring and autumn equinoxes, summer and winter solstices; perihelion and aphelion; angle of inclination of the Earth's axis of rotation describing a circle around the North Star; nutations (gear-like, figure 11.8) vacillating in a rotation about the Earth's axis. Each nutation covers about 18.6 years; a full cycle of nutations covers about 25,800 years, or a "Platonic year".

There is an incremental deviation of the alignment of the Earth's axis with each orbital cycle about the Sun. The rotation of the line through perihelion and aphelion rotates relative to the zodiac. The zodiac belt straddles the Earth's orbital plane, or Earth's solar plane. And there are the many motions of the Moon relative to the Earth and Sun.

Activities on the Earth, as related to the Sun and Moon, include: the sine wave or helical course of the Earth as it straddles its orbital path; the solar and lunar eclipses; lunar months; and the slow turning orientation of the twirling earth-moon barycentric arm; and other subtle notions.

External motions of the Earth are related to the constellations, that is, as the Earth's orbital plane relates to the zodiac. The precession of the equinoxes, in retrograde action of the Earth's inclining axis, reorients in relation the celestial spheres. And the tilt or planar overhead revolution about the major axis of the orbital ellipse, where the North Pole allegedly would turn over and assume the position of the South Pole and vice versa, this would totally changes the earthly view of the celestial sphere and not without affects reflecting back on to the Earth's geology, flora and fauna. In another possible celestial disturbance, the orbital plane would tilt its position, or turn on its ecliptic line, relative to the zodiac belt, while keeping its one focus upon the Sun, while remaining true to perihelion and aphelion.

In Hellenic Mythology, Helios's (Helius) son, Phaethon, took his father's chariot, figuratively the Sun, up into the Heavens, first too high and people shivered, and then too low and the fields were scorched. Phaethon could not control the four white horses. Are they the four "seasons"? Here is the anomalistic year's introduction to Man's history, as a shockingly noticeable phenomenon, with a Mythologic perspective on the advent of the seasonal weather cycles, affecting the four quarters of the Earth, its day and night halves, and its north and south halves.

The rotation of the orbital ellipse on its fixed plane about the Sun (focus) and the change in the magnitude and the eccentricity of the orbital ellipse are accepted phenomena, and they affect the winds and currents of the Earth, including distorting the "seasons" differentially and exaggerating, relative to the daytime and nighttime, and to the hot and cold cycles.

Interestingly, the Earth's angle of inclination may appear to vary with the orbital plane, but all may be the opposite, where the orbital plane tilts while the Earth's axis actually keeps to some particular point beyond the visible Heavens. The tilt of the poles describes each a great circle in either the northern and southern Heavens, perpendicular to the Earth's equatorial plane, which goes through the star Seirios.

As the orbital plane relates to the Sun, equatorial to Seirios, Moon to the waters, Mars to iron, etcetera, etcetera, could it be possible that each movement or element on Earth is governed by a particular heavenly body? Is it possible that accordingly the Earth is formulated and governed by each and all its formulators? And likewise, could the Earth have a hand in the formulating and governing of another or all the other heavenly bodies? Is this how all in the Heavens have been germinated? Is each heavenly body its brother's keeper, like Man on Earth ought to be?

It is common in the language of astronomers to talk of the heavenly constellations (zodiac) as moving into different positions, but for the most part, it is the other way around. The orbital plane of the Earth rotates and twirls at the same time the plane of the Earth's equator rotates along with the precession of the equinoxes, and along with other movements, relative to the celestial backdrop. The equatorial plane traverses the twelve constellations (as Man identifies them and numbers them) cutting through each of them at a different position from each twelve ages to each subsequent twelve ages to determine a kind of great sidereal year. And nevertheless, these twelve signs are basic for orientation in the celestial sphere. Although, too, the constellations move around with respect to each other, noticeable with the passing of multi-millennia, per Mythologies and folklore. Astronauts high up above see another view, that of the Heavens walking around in a new dimension.

Arguments for a solid Earth are often based on the marginal consistency of the angle of precession. That would be valid if the precession were looked upon as a deviating wobble of the sphere, but if viewed oppositely, that the orbital plane wobbles, the "precession" is merely of a trembling balloon matter. Especially with its mini-precessional nutations of approximately each sixteen years, such would proffer stronger arguments for a hollow Earth, For, a hollow Earth functions with significantly less inertia, with very sensitive reactions, as it quite frivolously bounces away from dangers and celestial accidents and repercussions.

— ——— —

If the heavenly world is a maximum image of the inner world, and if the atom is a compressed image of the universe, then perhaps difficult research can be extended on the basis of substituting the one for the other, as is done in simple algebra.

EARTH'S CALENDRICS

Coordinative timing integrates various terrestrial and celestial rhythmic cyclings and their ephemerally messing nodes.

49 — PRIMORDIAL CALENDAR

According to Mythologic and primordial traditions, the year contains 360 days. If in today's tropical year (solar or Gregorian year) we have almost 365-1/4 days, we are now spinning faster in each cycle in the Earth's orbit around the Sun. Or the magnitude in the area of the orbital ellipse has increased, if assuming the length of the day, proportional lengths of arc, speeds of orbital travel to be constants. A speculative converse argument is, if the latter be the case, there may have been a time when the Earth had 354 days (360 as the mean between 354 and 366) in its year. Since the rate of rotation of the Earth is erratically variable, there can be no constant year. Or possibly a year does not increase gradually in length of time, but until sufficient potential is accrued, it kicks over to another node of a meshing cycle (like shifting gears in an automobile).

Different civilizations had employed different methods and criteria for establishing the length of the year. Accordingly, each culture, having a particular fixation, or experience, had acknowledged a different type of year, such as the equinoctial, the Seiriac (or Julian), the sidereal, and the anomalistic, as well as other types of years based on the motions of certain planets. Of these calendars, only the Julian fits into periodic nodes with several other cycles, etcetera, to form a viable calendar.

Today's fixation on the equinoctial (tropical) year, with central theme on the equinoxes and solstices, bespeaks a "geocentric year", restricted to a solar relationship and to none other. It may be called a Sun worship calendar. It represents the instituted egocentricity of Western culture since the post-Renaissance. It ignores the sidereal universe and limits the telling of time to phenomena within the Earth's orbital cycle. This fixation is still a two-dimensional mental attitude, a paradoxic scientific sophistication built in the wake of the wavering foundations of a flat-earth society.

The length of the year has come down to us in rather confusing terms, not based on specifically defined kinds of years. The ancient Chinese year of about 2500 BC is given as 366 days, which may also lead to the suspicion of their using a primordial anomalistic year. The Persian and Indian years are given in the number of "days" fitting one 360-degree circuit around the zodiac, where one degree equals one "zodiac day". A zodiac day is independent of our days of light and nights of dark, but relates to the cult of astrology. A zodiac day offers the possibility that at one time in the past, a full day in reality had equaled one full "degree". The zodiac year (sidereal year), instead, may use actual solar days with intercalary days as necessary for earthly use. The Babylonian year may have reflected an ancient year of 360 days, which would fit into the 360 degree-days of a yearly circuit.

The hieratic Egyptian year is given as 365 days a year. Then, later, when a sudden change of time and circuit is noticed, when a bifurcation of time in the activity of agriculture occurs noticeably, the civil Egyptian year is given as 365.2500 days a year — that is, each fourth year has 366 days, instead of 365, a true "Olympiad", as is reflected in many seemingly unrelated ancient cultures. (Appearing arbitrary, or approximating, they fit with the various cycles within the great extent of the Julian Period, of 7,980 years.) The year of 365.25 days is found in many archeologic findings, besides the tropical. The ancient Sumerian year was composed of seven-day weeks and perhaps an intercalary week, or intercalary month of integral weeks, respectively and enigmatically giving years of 336, or 343, or 364 days, all divisible by seven. In contrast, in ancient Mexico, there is employed a year of the planet Venus, of 225 days, but its intermeshing cyclical effects perhaps are forgotten, other than being divisible by five. Note, many cultures had used five-day weeks.

The ancient Egyptians and Hellenes adhered to the year based on the heliacal rising of the star Seirios. Today, the equinoctial year promoted by Vatican is deceptively called a "solar" year and "solar calendar", and it is politically enforced throughout the world. There are African tribes, as the Dogon People, since antiquity still adhere to setting time by the heliacal rising of Seirios.

The four kinds of solar years and calendars are tantamount to historic facts. Do they depict changing timings, distances, conditions, and vacillations? Do the celestial dust and celestial water coming on through the exosphere, which keeps accumulating over the ages upon the Earth, cause terrestrial growth and continuously alter the terrestrial protective balances and the interaction of matter in our environment? Modern historians and scientists conveniently like to call these "inaccuracies" of measurements of "those" times.

Should ever in the ages to come the Solar and Seiriac planes (equatorial and orbital) converge into one, would the four kinds of years converge into a single kind of year, of 360 days?

– —— –

191

According to contemporary data, there are four kinds of true and valid solar years, varying from one to the other. In descending order in length of time, each of the following solar years is functionally described:

> *The anomalistic year: 365.2596 (±) solar days, 360.0032 (±) degrees of orbital circuit. It is for a hunting, fishing, nomadic, pastoral, and fishermen's life; for cycles of cold and warm winds, ocean currents with their varying directions, and fish and bird migrations; and it relates to surges of continental shifting. In AD 1200, perihelion fell collinearly with the "winter" solstice, and in AD 2000, it fell on 5 January. It runs past and beyond one full zodiac cycle each year, by 0.0032 degrees, cumulatively.*

> *The sidereal year: 365.2564 (+) days, with 360 degrees of orbit exactly. It alters mineral structures, facilitates astronomic investigation and measurements, and is used by the cult of astrologers. It is also known as the astral year. There is the longer "zodiac day", covering one degree of transit exactly, and is dissociated from the noon and midnight cycle.*

> *The Julian year: 365.25 solar days, with 359.9937 (+) degrees of orbital circuit. It is for the cycles of agriculture planting and harvesting,*

the astronomic synchronizing of cycles and nodes of cycles, and is a most ideal all around synchronic calendar. Hence, it is most ideal for religious holidays. It comes short of a full zodiac cycle each year by 0.0063 degrees, cumulatively.

The tropical year: 365.2422 (±) solar days, 359.9860 (±) degrees of orbital circuit, as of our century. But its measures elusively vacillate from century to century. It registers the longest and shortest days of the year, based on Earth and Sun interaction. It bears no synchronicity with any other heavenly or earthly cycles.

The Gregorian year: 365.2425 (±) solar days. Perhaps a fifth kind of year, attempting to define a fixed workable calendar for the tropical yeart. It is claimed as being the tropical year, but with the passing of centuries, its empiric formula for timing proves elusive and erratic. It bears no synchronicity with any other cycles. Vatican uses it to represent the tropical year, and for secular and religious settings.

Lunar year—there are several kinds: the year of twelve or thirteen natural lunations (Moons) applied to averaging out a luni-sidereal, a luni-Julian, or luni-tropical year. The correct Moon of the year directly affects planting and harvesting. The luni-Julian relationship governs the particular planting and harvesting days for each particular agricultural product. Called "months", they are also the civil or lengthened twelfths in any of the kinds of years.

The Julian year may also be known as the Seiriac year, or commonly said, the Year of the Dog Star, or of the Star Seirios, or

Sirius, meaning the "Scorcher". Seirios is the largest light-giving star in our galaxy; it disappears from view for seventy-two days each year and its year-to-year heliacal rising gives the length of the Julian year. The first full Moon after the heliacal rising of Seirios begins the new agricultural season.

The beginning of the sidereal year by traditional convention, 46 BC, is set at the first degree of Ares; this setting point is also called the "celestial spring equinox" (not to be confused with the terrestrial equinox).

The zodiac alignments of the Earth strictly govern the sidereal year and sidereal calendar, regardless of "seasonal" changes.

With the exception of the sidereal, to establish each kind of year, the precession of the angle of the Earth's axis of rotation governs in each of the formulas.

The anomalistic year, note overshoots the 360 degrees, the sidereal year is exactly of 360 degrees, while each of the other years does not reach to exactly 360 degrees in their full circuit.

The anomalistic and the tropical years may relate to geologic disturbances.

The lengths of the anomalistic, sidereal, tropical and Gregorian years in older references generally had been given up to eight decimal places in a race for more perfect accuracy. But by today it is realized that such is futile, because the years unpredictably vacillate in speed year to year, and so their lengths are rounded off more safely to four decimal places. Hence, in the above they are given to four decimals along with the plus-minus sign (±).

In general, all the dates of historic events are recalculated and given in terms of the Gregorian year, an unfortunate attempt for accuracy and uniformity.

The Muslim year is of twelve lunations, independently of "seasonal" changes.

Lengths of years to eight decimal places may serve in an ephemeris giving the length of each year for specialized usages.

— ——— —

Elusive to the imagination, there are several categories and kinds of months: the lunar (lunations), the zodiac twelfths (thirty-degree arcs), and the irregular twelfths of any of the several kinds of years. Further, a month's twelfth may be based a twelfth time of a year, a twelfth time of the number of degrees of orbit, and a twelfth of the orbital circumference of a twelfth of any particular kind of year, and many may be adjusted with an intercalary day.

Seven or more kinds of months are recognized by astronomers for the tropical (Gregorian) calendar alone, each month having its particular value, which would mean altogether there may well be over twenty-two kinds of months. The complete matter of months, however, is not germane to the matter at hand.

— ——— —

51 — NODES FITTING JULIAN YEARS

Cyclical time nodes are fittingly defined around the most ancient Egyptian, or Julian calendar, and occur as follows:

1 *Julian Period*
(per Joseph J. Scaliger);

15 *Great Paschal cycles*
(per Dionysius Exiguus),
(1st of year, 1st of month,
1st of week match in each 532 years);

105 *Callipic cycles;*

133 *indictional nodes*
(or 532 indictions);

285 *weekly nodes*
(each at 7 Olympiads,
of 1,461 weeks);

420 *Metonic cycles* (315 common,
and 105 short in each fourth),
of 532 indictions
(each of 15 years);

1,140 *sabbatical Julian years*
(each of seven years);

1,995 *Olympiads*
(Each 4 Julian years);

7,980 *years, Julian*
(5,985 common,
and 1,995 leap in
each fourth year);

98,700 *lunations*
(approximated);
323,855 *enneads*
(nine-day Roman weeks);
416,485 *weeks*
2,914,695 *days.*

1 *Julian-sidereal great year*
7 Julian Periods,
(of 55,860 Julian years, or
55,859 sidereal years,
that is, one less whole year
between the sidereal and Julian).

Years in the above outline are the Julian years! Or, may we call them "synchronic years".

Days are a period of noon to noon as experienced on Earth at any particular location; this period is divided into 24 equal hours. There are many other kinds of days, as viewed from other celestial positions and under other criteria; see in dictionaries or encyclopedias.

Lunations are natural months as viewed from the Earth, not "twelfths" of a year. The approximation comes to a fraction of one lunation within a Julian period, per contemporary measurements. There are many other kinds of civil and natural months; see in dictionaries or encyclopedias.

Great years are of several kinds. See in the binning of next Chapter,

All these These related frequencies in the universal Julian calendar do not occur in any of the other ephemeral calendars.

Note the frequent recurrence of odd numbers in the above!

The Star Seirios (Dog Star, Sirius) is the largest and brightest star in our Galaxy. The Earth's equatorial plane passes through the Star Seirios.

The Sun is the second largest star in our galaxy, The Earth's orbital plane passes through the Sun.

The Arctic Circle and the Antarctic Circle are the two parts of the Earth's surface, which, alternating, do not see the Sun for half a year, because of the Earth's tilt. One full day and one full night therein are equal to one full tropical year.

In calendric uses, fractional days are avoided. Intercalary days are employed rhythmically as necessary to bridge the gaps.

All four kinds of the "solar" years, as described, had been used by various ancient cultures, which in each case reveals the modus operandi of their cultures.

Copernicus and Scaliger, heroes of the astronomic and calendric sciences, knew well of the above relationships, as did the Byzantines and Ethiopians, a millennium before, and as did the Egyptians and Greeks in antiquity, before the arbitrary change in the West. Therefore Copernicus and Scaliger vociferously and strenuously, risking the stake and fire, had objected to a Papal change of the calendar in 1582.

— —— —

52 — PLATONIC AND GREAT YEARS

The difference in the length of time between any two kinds of years determines each of the several kinds of "great years". Out of the four basic kinds of years, six kinds of great years are categorized. Each kind of year covers a different arc length of circuit. Hence, a different timing of its cycle, and a different arc radian measure for each orbital cycle, albeit they all are along the same orbital path. That is, there is a slight slippage in each year between each pair of years. When the incremental slippages come to one year, relative to the other kind of year, throughout the many years reach to a difference of one whole year, then we have a great year. The number of years of slippage between the kinds of years is per measurements valid today. Such may vacillate from age to age; therefore, they are rounded off to four decimal places, per current practice. (Not too long ago, references were striving for nine-point accuracy, but in vain for the unending seemingly non-rhythmic undulations.)

The difference by one year between the pairings yields the following:

> *Sothic Period: in each 1460 full, or sacred Egyptian, or Sothic years (like Julian years), there are 1461 primordial, or vague Egyptian years.*

> *Anomalisto-sidereal great year: in each 111,300 anomalistic years, there are 111,301 sidereal years. Perihelion moves counterclockwise across each three zodiac signs in each 27,825 years. In 55,650 years perihelion is halfway around the zodiac in this AS great year.*

Anomalisto-Julian great year: in each 37,900 anomalistic years, there are 37,901 Julian years. Rise of Seirios moves from major axis to minor axis in each quarter of the elliptic circuit in each 9,475 years, or quarter-way around the ellipse. In each 18,950 years the heliacal rise of Seirios moves half way around this AJ great year.

Anomalisto-tropical great year: in each 20,900 anomalistic years, there are 20,901 tropical years. Perihelion moves from solstice to equinox in each 5,225 years, and from equinox to the other solstice in the next 5,225 years. In each 10,450 years the two equinoxes and the two solstices respectively reverse their positions halfway around this AT great year.

Sidero-Julian great year: in each 57,400 sidereal years, there are 57,401 Julian years. Zodiac moves three signs to any fixed date in each 14,350 years. In each 28,700 years, in each six signs, or halfway around this SJ great year.

Sidero-tropical great year: in each 25,800 sidereal years, there are 25,801 tropical years, commonly called Platonic year. Zodiac moves from solstice to equinox in each 6,450 years, then equinox to solstice in another 6,450 years; and the hottest day in the year happens to fall on one equinox, and the coldest on the other equinox. In each 12,900 years halfway around the Platonic year, or ST great year.

The Juliano-tropical great year: in each 46,800 Julian years, there are 46,801 tropical years. Rise of Seirios moves from solstice to equinox in each 11,700 years, as sowing and harvesting seasons fall

behind from solstice to equinox, and from equinox
to solstice in another 11,700 years. In each 23,400
years halfway around this JT great year.

Each kind of year lags the other, each with a different length of arc radians to accomplish its intention. Each has its differentially peculiar stresses, strains, temperature, and pressure cycles on the Earth's motions and on its flora and fauna. So, the question is, if longest and shortest day of the year are not related to the hottest and coldest days, what is a "season"? Each is a complex phenomenon cyclically changeable and of necessity re-definable, relative to each respective age.

(See *The Orphika,* by Ioannes D. Passas, with foreword by Constantine S. Chasapes, astronomer, in Greek; also *Julian Calendar Valid, by* this author.) This observation eventually will nullify the arbitrary "sacredness" attributed to the geocentric tropical year, as we know it, or the sun worship year.

In each 240,000,000 our solar system makes one revolution in our galaxy,

– —— –

53 — HOURS OF THE DAY

The mechanical clock, an invention of the last millennium from Byzantine Constantinople, ticking steadily and rhythmically, divides the day into twenty-four equal hours, each of equal duration of time; that is, most unlike with the timings of the "seasons", the sunrises, and the movements of shadows. Upon this invention, our concept of what is an hour is changed. It is completely divorced from the sunrise and the sunset, and the shadows they cast. Now the mechanical hours are given as occurring at a different time from each natural morning and each natural evening, accordingly. The hour is freed from the noticeably short half day from sunrise to noon and from noon to sunset of "winter", and from the long half day from sunrise to noon and from noon to sunset of "summer". It has given cause to look upon the day as composed of twenty-four absolute hours with minutes of equal duration and seconds of equal duration. By convention the day of twenty-four equal hours is set to begin at midnight to midnight of each twenty-four-hour period, in total disregard to the natural timing for sunrise and sunset.

When we say "hour of the day" and "time of day," they are not synonymous. The length of time, before the coming of the ticking clock, is related and evinced in its seasonal changes. Each hour of the day varies in length from hour to hour. The length of the shadows from day to day varies through the yearly seasons in contrast to the simplicity of the clock's absolute hours with its steady metronomic ticking.

When time was told by the Sun's position and the stars's positions, accordingly the hours incrementally grew in length of time toward noon and waned toward midnight in "summertime". Oppositely, the hours waned in length, incrementally shortening their time toward noon, passing noon, they grew long toward midnight in "wintertime".

The hours of the night are told by the positions of certain stars, as in wintertime certain lower stars begin disappearing until the solstice and then begin again to rise into view.

A more sophisticated sundial has the hours differentially laid out in several curves, graph-like, each curve for the particular parts of the year, to adjust for the "seasons", proportionally to fit the changing lengths of the hours.

As either of the equinoxes approaches, the hours incrementally become equal in length of time, and on each equinox, all twenty-four hours are equal in length of time and match the time as given by the mechanical clock. The average measure from day to day always is given at high noon, the fixed fulcrum for the hourly changes. The sixth hour of the day and the sixth hour of night are the same as 12:01 PM and 12:01 AM. The ninth hour between noon and sunset, as in our epoch, in the summertime is closer to 4:00 PM, and in wintertime it is closer to 2:00 PM.

(See figure 53.1, superimposition of hours: equal during equinox and gradually varying during the rest of the two half years; the outer ring of numbers are of the mechanical clock.)

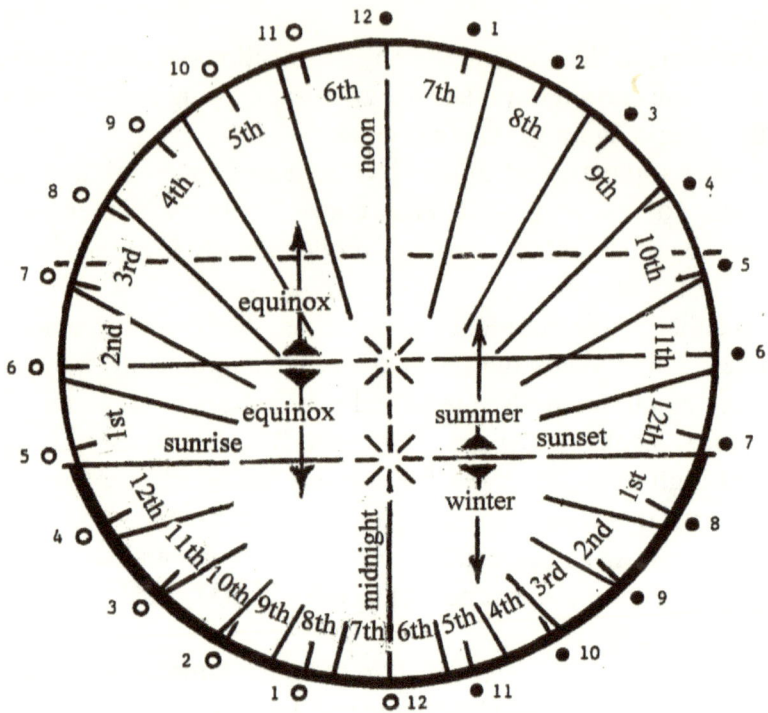

***53.1 — Diagram of the waxing and waning hours of the
day contrasts these to the perennial equal hours of the
mechanical clock.***

The twenty-four hours of the day-night period, above, when
projected from the center of a circle in equal angles of fifteen degrees
apart, represent the twenty-four fixed mean or standard solar hours,
shown with dots. In the scheme, the twenty-four hours are projected
from a point variably below the center of the inscribed circle for the
longest day of the year (today commonly called the "summer" half)
and above for the shortest day of the year (today commonly called
the "winter" half). In both cases the ecliptic adjusts as it move
incrementally for each day between the bottom-most to the top-most
for the Earth's ecliptic, while keeping to the twelve solar hours of
the day, and to the twelve hours of the night (24 x 15 = 360 degrees).
The fifteen-degree parts radiating from the off-center position in the
above diagram represent the waxing and waning temporary solar

hours; these are shown by the arc of each fifteen degrees, as each arc radian lengthens and shortens. Noon and midnight of the standard hours of the clock and the temporary hours of the day and of the night are always in concurrence.

The use of the "hour of the day" and "hour of the night" are important for the healthy digestion of food, taking medications, sowing, harvesting, planting forests, lumbering, and infant care. Knowing the ancient rules produces tastier foods; healthier growth of human beings and farm animals; longer lasting woods for construction, furniture, and implements; and provides for the better digestion of victuals. Watch how the animals live in their natural habitat, defying Man's "o'clock" and instinctively reacting with nature's time of day.

The use of the "time o'clock" is ideal for fiscal and mechanical uses, scheduling. However, seasonally setting the clocks off by an hour is doubly shocking to a human being's biological timing, because a natural transition is tastelessly being countered.

— ————— —

EARTH'S
SYNCHRONICITY

Understood and non-understood phenomena, nevertheless
perpetually, harmonically and beneficently work together.

54 — REALITY OF ELUSIVE SEASONS

An earthly "season" is composed of four unrelated basic factors: hot or cold days; long or short days; calm or violent winds and waves; humidity and aridness; and particular astral settings with their variety of physical effects. Commonly, but mistakenly, we take all these as a single given for determining a "year". However, each of these basic factors, each in its own cycle, incrementally bifurcates from the rest. Confusion is avoided by coming to understand the specific function and specific timing of each.

In review, the nature of flora and fauna is altered from region to region in the geography of the world, and from age to age. China, at about 2500 BC, was hot and humid, the Sahara was grasslands, and central Europe was covered with glaciers. In ancient writings the star Seirios had a distinctively scorching bright red color, perhaps with significantly noticeable affect on the Earth's flora and fauna. However today it is the most brilliant white star in the Heavens, and still uniquely and strongly affecting the Earth, although not as apparently. Could the red have been intrinsic to something, yet to be discovered?

A brief review on the bifurcating of the phenomena, which within this age in the span of multi-millennia, serves to define "seasons": The nature of the "seasons" is elusive to our present "idée fixe" of what is a "season" and to the veracity of today's dictionaries, which lean toward current semantics and away from developmental etymologies. ("Winters" are cold with short days is an ephemeral description, true only in our present age.)

The Northern Hemisphere in our era leans away from the Sun, as the Earth whiplashes closest to the Sun, at perihelion. In AD 1200, the world experienced the mildest all around climate within the historic age. The times of the solstices and the major axis of the orbital ellipsoid fell collinearly, of winter solstice with perihelion, and of summer solstice with aphelion. It is when Europe and Greenland experienced the mildest climate, when Greenland was hosting farming and cattle ranching. And civilization spread into the northern lands of Russia. Today perihelion lags behind the solstices by two weeks. These facts contribute to destabilizing and altering the seemingly accurate calculations over the centuries.

— ——— —

55 — MEASURES ALL SPECULATIVE

The units for measuring time and space can be just as suddenly changed into another scale, with their sequential relativity being totally upset and rendered non-intelligible. Beyond six thousand or six million years ago, for instance, the units we use today may be registering a year as a million years, or an hour. A meter or yardstick may equal hundreds of miles or a fraction of an inch. The weights may be a thousand-fold amplified or minisculized, leaving the changes in the timings and lengths and weights to be, at least, of endlessly changing relativity to the each other in each of the ages.

The Egyptian year suddenly had jumped from 365 to 365.25 days, with no record of transition. The Chinese ancient year of 366 days with no intercalary formulas and no transition leaves us with another mystery.

Although in our times such concept remains quite incomprehensible, such as in: the world was created in six days. Adam and Eve live about seven thousand years ago, Man lived for nine hundred years, etcetera, and truly the author believes that we do not understand these records, rather than to brand them as non-scientific. Were these six days of very slow time cycles? Or, were they "celestial days" and "celestial nights", in contrast to the terrestrial days and nights, which are creations during the celestial fourth day (a concept not yet understood today)?

The relativity of time may not be limited by any means to just a slow "evolutionary" and "orderly" process; time can be erratic, and it can be momentary; as a "day equals a year," a day is the time needed for a moral, physiological, sociological, or religious (per theologians) accomplishment. One ought to be more tolerant in the matter of years and centuries per the Old Testament and ancient lore, and per today's scientific fixings, because "fixings" defy the plasticity of time. Imagine a graph of square inches printed on a balloon being inflated with air, enlarging the squares ever-distorting them, until *poof.*

Volcanic action, earthquakes and lightening bolts can be very sudden, in contrast to "evolutionary" movements. Likewise, just as suddenly and abruptly anything can happen, examples are: the marginally offsetting of the angle of precession of the Earth's axis; the climates changings; the breaking up of Pangaea; the "forty days" flooding of the "face" of the Earth; the popping up of continental plates and the rising formations of mountains; the sudden accumulation of ice over the poles, and the melting of glaciers with a sudden change of temperature being distributed throughout the Earth, with the creating of new rivers and lakes.

As for flooding the "face of the earth", does that mean flooding a portion of the Earth called "Face," which hosts the civilization of the line of Seth, third son of Adam and Eve? Or does it mean the entire surface area throughout the Earth? If "face" is limited to legitimate civilization in the eyes of the God of the Old Testament, perhaps the descendents of Cain and Abel remain not affected by the flood. Cain's descendents would be the Oriental and American Indian peoples, and Abel's would be the African peoples, both of which took to living beyond the "face" of the Earth.

All this can happen without total destruction, but may be taken as compensational changes in the balances of nature, or allowance of the Supreme Intelligence, or a reorganization of mankind. Calendars for the telling of time for the days, weeks, lunations, harvesting seasons, wind, ocean currents etcetera, each of such answers to a particular kind of year, and all, jointly or severally, can just as suddenly be caused to change. All can just as suddenly change relative to each other, breaking from their fixed differences. Supposing there was a time that the Earth made only two revolutions while orbiting around the Sun, instead of 360, or 365, then a year equaled only one day with one nighttime. Can our chronometric gadgets and carbons pick up these metered differences out of the reports lodged by archeological researchers?

How coldly cruel does all this sound to helpless loving human beings? Nevertheless, and not fully understanding God's allowance since Man's creation, the animals sense such imminent dangers and try to tell us, but are not trained to talk correct English. The animals communicate with us only in their language, including their body language; they publicly convey, all that is about to happen. But who wants to pay attention, if they are only inferior animals and don't speak our language?

Units of all kinds of measures may be inflated and deflated to fit new criteria, if not by following changing criteria. Just like the value of fiat money, which intrinsically is worthless, being simple paper with numbers printed thereon (poor trees and octopuses), which the politicians receive from the private usurious foreign banks, paying heavy interest to them from the day of their being ordered (borrowing from them), even before this fiat money goes into circulation (which is a gross violation of business contracts). Then they put such fiat paper into the hands of peoples, for profits in interest, for profits on our debts, profits on our losses, profits on penalties against us, and even profits on our meager gains. So, then, why cannot a century be a day?

— ——— —

It seems stylish for lexicons and encyclopaedias to talk more and say less on certain subjects. References before World War One give full and sincere answers, on measures of astronomic phenomena. Especially the matter of perihelion–aphelion is targeted for dumbing down, as well in most other recent publications. Only the mechanical definitions of perihelion and aphelion are given, but fail to give even any distances, leaving one to wonder.

One of the most prestigious university publications in England, 1990s, the Oxford Encyclopedia of Astronomy, under the entry 'perihelion', typically gives the mechanical definition, but fails to give even an approximate distance between Sun and Earth. But under its entry 'aphelion', of lesser importance to researchers, it does include that distance. No doubt this editorial slip proves a behind-the-scenes deliberate intention for dumbing down. But why target perihelion? Is the eccentricity of its elliptic path a key to the facts of global warming?

— ——— —

56 — FLICK OF THE EYE

Simple theory: The Earth's shape vacillates over millennia between extremes from tangerine to lemon shape. But because of the Earth's more complexly varying intermediate motions, the shape may approach that of an apple or bell pepper, watermelon or peanut (from increased wobble, acorn, turnip, parsnip, etcetera; a shape may by symmetric, "lopsided", a torus, or temporarily without form. A shape may never return to its exact original configuration. The angle of precession may approach ninety degrees to lie flat on the orbital plane. Any of the shapes might adept primarily by the water masses acting against the landmasses, or vice versa.

This means that dry lands with civilizations may disappear under water, and new virgin lands and new mountains may appear at any given parallel on either side of the equator or at any meridian. This means that climatic conditions can change outside of any reasonable pattern or prediction, which is commonly called in the insurance business "acts of God". These may well be prophetic. In this context we ought to chuckle, when we say: "See my piece of land for when I retire?" Or, "I own a quarter of this equatorial jungle to escape cruel civilization."

Compare, for instance: today the mountain ranges of the "Western world" mostly are aligned north and south, while in the "older world" they are mostly east-west and diagonal. In the Western world, there are no geographic obstructions for winds traveling between the equator and the poles, giving unobstructed cause to the sudden rise or drop in temperatures, and delivering a shock to civilization. In the "older world," the several east-west mountain ranges tend to stratify climatic zones and allow for greater predictability for planning for an assured perpetuity of civilization, and environmental readjustments in general.

Question: Besides the forces of the cosmic layers, can the "vibrations" of social positivism (as in group prayer, social laughter, mercy, brotherly love), or of social negativism (as in curses, indifference, revenge, non-satisfaction of the law, hate), have a serious reaction upon the Earth's crust, shell, and atmosphere? Could all such culturally be sustained over a period of time or for the moment, whether by the entire population, a region, or a powerful few. Can group petitions, uniformly loud and clear, ward off a "catastrophe", or induce "divine" happenings and good opportunities? Many traditions say, yes.

Is it possible that at one time, when the orbital plane and the equatorial plane were the same, all the solar planets had their abode upon this one and same plane? May we call such the "great galaxial or solar plane," somewhat mindful of Saturn and its satellites? With all planets lying on the same great plane, would interplanetary travel be made easy?

Since the equatorial plane goes through the Sun, is it possible that at a primordial time the Earth's orbit was around the star Seirios (Sirius) and then it jumped over to orbit around the star Sun? And later the Moon also jumping orbit rejoins the Earth now orbiting about the Sun together?

Is it possible that the two stars may shift vertically with respect to each other and some other third celestial reference, resulting in the Earth's tenaciously adhering to both given celestial relationships, and with a third relative plane or point? Or may it be said, the Earth and Sun, both orbiting about Seirios, at a critical juncture the Earth and Sun, together forming a spinning Sun-Earth duet are spun off from another galaxy, to become a satellite of Seirios, along with the creating of a new orbital plane?

So, with Seirios tugging on the Earth's equator and the Sun pulling on the Earth's axis of rotation, then the Earth's relationship to both of these stars should be characterized as a "universal joint", and possibly the pair should relate to another pivotal reference yet to be discovered and identified. And, can the magnetic axis relate yet as to another historical pivotal reference? When the Earth's relationship is stronger to Seirios, the Earth may take on the form of a tangerine; and when with the Sun, a lemon.

Allegorically, the three verb tenses —past, present, and future— together may be looked at as either in a spiritual, or in a material sense.

Spiritually, is when all is in the present tense in the ongoing and all-inclusive time of the transcendental "Day One" (per Genesis), with the future and the past being situationally relative to the all-present flow, to the ever-present principle.

Materially is when the past and future are fictional; rather when they are obstructive to the instantaneous present of practicality, or of negotiability. The "present" relates to the five senses of Man, to react to whatever for survival or profit, and accordingly the five senses may relate to the duration of the present tense, and all beyond is but speculation.

To a moderate person, the three tenses are signs of the particular times on certain matters. Although it is up to speculation, as to how broad is the present tense, from a material, legal and moral view. The past and present serve as a vector to project a particular expectation of the future. Or by seeing the future and knowing the present, a probability on a past condition or fact may be evaluated.

— ⸺ —

Man with his potential ingenuity may strategically construct bridging pathways between firm "clouds" or certain nodal points high in the atmosphere, or higher. First, energy lines or beams are established, like starting the construction of a suspension bridge with two strands of wire, securing these between the nodal points as indicated by nature. Then these energy lines between are particularly energized to attract certain physical matter and to coagulate the accrued matter into tough negotiable cables of high tensile strength (but cautiously mindful of the suspension cables in the tragic story of the "Bridge of San Luis Rey"). Then proceeds a system of harnessing and converting energy into substantial matter, into platforms and galleries, islands, and even into soaring continents. Celestial highways may be established by permanently beamed energy currents, which may serve as invisible runways, lanes, or tracks for celestial vehicles (similar to the new railroad systems now being built on Earth, whereby trains follow along on a given path of Eddy currents without touching the rails or roadbed on the ground) to link the celestial islands, etcetera, for travel and trade.

On these, as is already known about Eddy and other currents, solid construction may well be floated upon such currents up in the heavens, just as heavy battleships at sea float upon water.

Celestial dust, and more solid celestial particles and "garbage" may coagulate to fill in certain celestial patches to raise food products, and for animal husbandry, etcetera. Thus we would have a double-decker Earth, or an Earth with suspended "solid" heavenly galleries. Airplanes, today, are crisscrossing the skies weaving visible chemical trails for "unknown" reasons; the use of airplanes accordingly, may facilitate coagulating these into useful air-passages, waterways and celestial energy conduits.

As for water up there, that, too, is possible. Supposedly the spinning Earth below somehow changes its rate of rotation without altering its particular spheroidal shape. The result would be more volumes of water would be evaporated, partially shielded from the Sun, and be attracted to reach higher to release rain upon the thirsty supra-terrestrial islands and continents. The evaporation of water may be out upward from the equatorial belt or parallel upward adjacent to the polar axis and parachuting outward, depending on the faster or slower speed of the Earth's rotational change.

Would this help Man's problem of over population?

— ——— —

Now, let our imagination go inside the world. Supposing a non-organic asteroid or missile, or an intelligent evil, penetrates the Earth's atmospheres and shatters a good portion of its continental surface. Is the Earth destroyed? Most likely, No! The invisible spun shells of energies engulfing the Earth may well have their counterpart of shelled energies above and below the Earth's shells, lying between the inner terrestrial surface and the little sun. And like all the interiors of atoms, the Earth may exhibit a resistance of unimaginably higher durometer to penetration than Man may yet conceive of, and to hold the mass of continental shambles in place, pending nature's structuring new continents, if not restructuring, for other energies are in the play.

As well, if on notice of an imminent danger approaching the Earth, the new academic breed of celestial engineers may quickly arrange for the thwarting of the undesired (not unlike in today's "sandbagging" to hold back a flood.)

Further, if Man uses cables, nodes, platforms and bridge-ways composed of energies for the bases of structures in the sky, then upon grasping the principals of the energies within the hollow of the Earth, could he not smoothly and with little effort combine these combinations to reinforce the inner shell of forces to a higher degree, to a tighter web of energies, and as well to control the outside pressures in the remolding (akin to genetic engineering) of the Earth to a more ideal shape? (A dolphin keeps readjusting its body contour according to its speed, depth and other environmental factors,) And would not our Planet Earth then become a pioneering "model planet" for better and happier living in our Galaxy?

And would, then, the Bankers not be needed any more in such world of advanced higher technology and deeper initial Classic thinking, while requiring ever so less capital means to get started? Technology by then would be a bargain, as like a little girl picking golden-yellow daisies in her backyard to reward her grandparents for giving her all the attention that is needed. Bees build their beehives for honey, beavers build their damns altering rivers, ants build their hills with chambers and tunnels to store food and procreate, boys build their tree houses for the pleasure of accomplishment, etcetera, all without having been educated to think first of paying a visit to a money-loaded Bank. For what, unless for bubble gum, movies and ping pong table rentals?

It makes sense that the gravitational fields and the electromagnetic fields are not the same. (See, per Timothy Leonakis, in the ADDENDUM, to whom the undersigned is indebted. As he presents it, one more simple argument may be deduced for a naturally non-collapsible Earth Hollow.)

Natural electric energy existing in the atmosphere ready for tapping, eventually when known how to be tapped, is enough to supply the entire Earth in perpetuity, without need for manufacturing energy. (*The Field*, by Lynne M. Taggert, is interesting reading on unexpected tremendously powerful energies in small spaces (as in a two ounce decanter) to make a house comfortable the year round, and on speeds and communications faster than the speed of light.)

This opens to other horizons, to new simple lightweight hats, clothing and shoes, which can envelope us with good comfort, indoors and outdoors, through cold winter and hot summer, whether in an enclosed vehicle or on a bicycle. Such energy could warm or cool our clothing, or they can generate a comfortable aura all around us. This means as well to be able to sail through outer space on a raft or skiff, or along celestial bridges, ways and rails composed of harnessed waves of energies, and to and from other planets as Mythology suggests.

— ———— —

Is Man stepping out of bounds by staring his dreams into the sky? "Let there be a *firmament* in the midst of the waters, and let it divide the waters from the waters." *Firmament* etymologically is a steadying support, or fortification. Some wrongly translate "*stereoma*" as "space", a vacuum. But *firmament* or *stereoma, in the Septuagint, in Genesis 1:6,* implies a domain of structured existence and activity, but not an idling emptiness. Moses, the redactor of Genesis may be suggesting to us, that God would not have objections, should we go off staring on high, once we are able safely to proceed doing so.

57 — DAY ONE OF CREATION

The book of Genesis speaks of a "Day One" of creation, and proceeds to discuss the second, third, fourth, fifth and sixth days of creation. Note, it does not say, first day, then day two, day three, etcetera. This is most symbolic, as "Day One", a cardinal number, is yet ongoing in it's being perfected and yet incomplete until today. Now, whether Day One gently had come into being, or suddenly, is a conundrum, for it is a happening from before the coming to be of the human conscience, therefore practically a moot point as for the history of Man. Therefore, the physical history of the created universe, an inexhaustible matter, herein is lightly touched upon.

And within this on-going "one day", there occur the second, third, fourth, fifth and sixth days, in ordinal numbers. All is happening within the duration of this ongoing Day One.

In Genesis 2:3–4: "Let there be light; And God saw the light, that it is good; and God *divided apart* between the light and between the darkness. And God called the light day, and the darkness He called night. And there was evening and there was morning, Day One." (From Septuagint Old Testament.) The darkness in this case, it may be presumed, is absolute darkness where nothing can be seen; it is not the relative darkness on Earth where we still can see in nighttime. Darkness, perhaps, pertains to black hole, a mass impervious to light. Obviously, in Genesis this early day and night pertains to a celestial kind of day and night, as it will be not until the fourth day, when God will create the two great lights for the particular day and night as perceived on our Earth. Note, God *divides apart*, between the light and the dark; it is as though they are of a comparative category, that is, night is not merely the absence of light, but of some other essential fulfilling energy, or substance (see Apostolos Makrakes). (And see in Chapter 36, on icosahedrons.)

In the Book of Genesis, the creation of fish, reptiles, birds and animals is given in the plural. But Man is in the singular. The significance is that Mankind is mandated to be as a single democratic body politic, a uniformly united corporate culture of structured republican divisions, and each person the keeper of his brothers, all in a oneness. Animals and lower beings are physical individuals, which upon maturing no longer need parents, and remember them not. Hence, the author capitalizes the word Man. The New World Orderists are aware of this and strive for their "takeover" of the world as soon as possible, before the awakening of the Peoples seeking their "oneness" democratically from bottom up, as though only the initiated elite have eaten of the Apple of Knowledge and of the Apple of Life to set all the world aright.

— ———— —

At the conclusion of each ordinal day (excluding Day One): "And God saw that it is good." This should mean, He has already set the energies, mechanisms and the laws of compensation for the "rolling of each day", and that He oversees the outcome, bespeaking His foresight, conscience, love and truth on behalf of His ultimate Creation, Man. But an Atheist, already a god per his ego, considers most all other religions beneath his dignity.

God, that is, sees, that is, as He intends, the light He created is an ongoing and working reflection of His Intelligence. And the night is an intrinsically on-going perfective medium for the silent activity of growth, or expansion, again as He intends, while Mankind is asleep at rest, pointing toward the ultimate of perfection. Day and night are as a revolving and perfecting duet, while always under the view of the Sun; therefore day and night are not discontinuous and separating, rather actually sequential and connecting. And He intends that Man keep up with the course of His ever-waxing Creation. Not by new fads requiring destructive changes, but He charges Man to keep improving and repairing "all that is" in his ever-expanding domain, into whom He had inbreathed His Spirit, to be keeping "all that is" in the state of "goodness" and in a condition of "cleanliness", in each day.

— ——— —

The ancient Greek philosophic schools, as well the Book of Genesis, see "evolution" as cyclical, let us say, as a helix. In other words, the five days, second, third, fourth, fifth and sixth, of Genesis are veritably marked off rolling over periodizations, distinguished by the "nights" between them, by a "night" node, marking the accomplishment and completion of each day, pleasing to the Creator, all within the yet ongoing Day One, that is, within the potential limit of "days", perhaps, until "It is finished!"

And we mortals, as is ordained for us to discover and understand, find ourselves domiciled somewhere between the infinite reaches and the infinitesimal reaches of our incomprehensively scaled universe.

Man fits, or should fit, into creation physically and spiritually, because mentally and spiritually he is able to, and does transcend both into the superior and the inferior stages of his physical environment. Conceptually Man is aware of the unknown. And God said on the Sixth Day, "Let Us create Man in Our Image and in Our Likeness." He did create us in His Image, but as for potentially His Likeness, Man, each willingly by his meritorious participation and cultivation toward forming a single body of Mankind, and in the time ordained, only then can he accomplish the Likeness.

— ——— —

Some of the dictionary definitions of Myth pertain to fairy tale; such definitions the undersigned author rejects, as prejudicially improper, or at least as late irrelevant interpretive developments. The ancient Mythology has nothing to do with matter pertaining to Hollywood and the comic strips. Myths are true episodes, good and bad, of moral conscience and in violation, transmitted into history in laconic symbolic language, which provoke and need interpretation, so to discover a cornucopia of intelligence lost history, which includes of the entire celestial sphere. Matters in Mythology appearing to be as Pagan theology, may not necessarily be so, but there may be some preemptive actions and meanings in a different verbal tense, intending to be prophetic, as virgin births, as fatherless heroes, as born from a god's forehead, etcetera. Mythology in the early Christian period is clandestinely made into an underhanded political weapon to attack Christianity, hence, unfortunately, its bad name.

However, interesting, in the Nicene Creed the phrase occurs, "begotten of the Father *before all the ages*... not made". This means, it neither ignores nor contests the Pagan Mythologists from the past series of ages, but gives them succinct answer, that the "before all ages" means even before the "Day One".

If we were to allow ourselves into a Mythologic frame of mind, as we see the internal terrestrial convulsions and the celestial convulsions, we may readily see our own hygienic, personality complexes, and spiritual convulsions as well. For Man is in the image of all creation, as well. Are all these convulsions food for repentance, whether in politics, education, morality or Christian perfection?

Keeping in mind the vital enigmata being succinctly divided between Mythology, on the one hand, as being of essential record (history) of the habits and transgressions, progress and failures, and good or bad, for supplementing Mankind's conscience, and, on the other hand, as revealed religion, as a means of Salvation, the author as well capitalizes the word Mythology.

— —— —

Hercules did not burn down the Augean Stables with the ages of accumulations therein, as being dirty, smelly, obscene and hopelessly outdated; rather, he diverted a river through them, washing them out, clean of all those negative accumulations over the ages, and therewith fertilizing the life-giving fields downstream.

This Herculean story is allegoric to those demanding "changes" in society. "Change and Changes", as is used in modern "Liberalistic" language, means violently to destroy and whimsically to build anew per ephemeral fads, not distinguishing matter from manure. Because every ten years, semantics and all else that naturally is ephemeral undergoes distortion, "changes", as in fads, sense of aesthetics, conformance values. In his currently spoken language, semantics run amuck and away from the etymology of words and phrases. So, when Man talks about the doctrine of "Changes", often he does not know what he is talking about, he does not know the words he is using, and neither do his listeners, even when, or especially when talking at a fifth grade level, as though merely approving of each other's poetry. Languages, too, dilapidate and head for destruction, unless disciplined into the original intent of each word.

Let us picture the voluminous reports on all the needed changes in society, piling up in a warehouse, decade after decade, and losing their relativity, before any such report manages to be read and heard, with the consequence of failing to bring on some radical destruction. A river of clean water must be directed through the piled up and outdated storages of ephemeral reports on recommended social "destructionism" of mountains and oceans, for recycling. The true changes upon what Man has wrought, is in his house cleaning, and his getting matters back in order. Man cannot change the world according to his personal godship, but he can heed to the duty to clean up. And only in this way can Man progress and progress, and not be spinning his wheels, or falling backward by back stepping.

— ——— —

A Buddhist theory is that chunks of ice are constantly being rained into our atmosphere from outer space, and they shatter into mist and rain to precipitate upon the Earth. If that were true, combined with the celestial dust being trapped into our atmosphere, the mass of the Earth, and all that involves the entire Galaxy, is in a constant state of being nourished and growing, that is, all created existence physically is still in the Day One. (See *The Hidden Message in Water*, by Masaru Emoto, 2004.)

— ——— —

EARTH'S MINI WORLDS

Earth's inner and outer dynamics and gifts contribute to the feautifully sustained liveliness within and without.

58 — AN ATOMIC ADVENTURE

An atom of any physical matter, any element, is contained in a series of succeeding envelopes, as molecules, cells, tissues, organs, or layers, similarly as all contained in the galaxy, and on down to the exospheric layers in the space over the Earth, downward to the Earth's surface and continuing beneath the Earth's surface. The "atomic activity" within each atom is at speeds perhaps stupendously greater than the speed of light, perhaps at speeds approaching the human mind. For these tiny forces within each atom must withstand the combined forces of the whole of the universe in order to maintain their atom's integrity.

The little atom formidably resists destruction from outside influences, no differently than the checks and balances of the Earth's resisting destruction, while suspended in the Heavens. The atom theoretically can be dissolved and reconstituted into a different form of matter through such high speeds and minute "electric" forces, which may overpower in magnitude other atoms around it, and within the atom. Or the atom can be reshaped into potentially having surprisingly altered functional characteristics.

The Earth's functional characteristics, it is shown, are altered through the ages in its changing configurations, perhaps cyclically, as in the weather, the length of calendric units, the kinds of biologic and botanic life it sustains, and fertility. Since matter is deemed indestructible, the Earth or the entire solar system can be reconstituted into something else, be it by any greater indescribable forces, besides per chance by meteorites and "planets crashing". Scary? Man's soul is indestructible too, which means Man must always be prepared for his eviction from this world, perhaps to the metaphysical, if not to another galaxy, and if not to where he does not want to go for fear. Scary? Man may not escape into his own ignorance.

So far, in trying to enter an atom, to tap and harness its forces, or to influence an atom to behave a certain way, Man has only been able to make it explode, to destroy its present nature, like piercing a balloon. He is not yet able to find an "opening" to make his entrance to an atom, peacefully and usefully, like modern surgery by which no longer cutting open through the muscles is needed.

— ——— —

It is an ongoing paradox that educators resort to such teaching devices, as in a museum, as taking a tour through the inside of an atom —in an exaggerated size of a mock-up of the atom— around all its defined and color-coded fields and internal spheres, through intriguing little stairways and cutely curving bridges for fully grown visitors and their children. It is a further paradox that no suggestion is offered on how this cell had been properly entered, whether through polar openings or dried up oil veins, or what. Still, however, a true walk around the inside of the Earth is impossible for "them" to conceive, perhaps with the phantasmagoric exception of the genius Jules Verne.

Could the atom be a miniature Earth, with all its series of dynamic internal and external envelopes? Or, further, could it be a mini galactic system? Can the concept of electrons within an atom jumping their orbit on to another atom find parallelity of planets jumping their orbit on to another star?

— ——— —

59 — PANGAEA RECORDS REVISITED

Ancient maps and records, which depart from the norms of contemporary geographic knowledge, generally are taken to be somewhat frivolous, as out of proportion, and instead the modern cartographers make present-day credence to be retroactive, as though it could never have been as the ancients had delineated them. They present such old maps self-servingly to demonstrate how advanced we are in our modern "greater intelligence" in the field of science. Allowing for some hearsay and imagination of some older and ancient cartographers, perhaps these ancient maps depict more truth and changes actually having taken place. One of the great examples of such is the map of Krates of Mallos, Hellenic, 145 BC. It divides the globe neatly into four quarters, that is, four continents, which are separated by two oceanic belts, the two forming a cross. the Aequatorial Ocean, straddling the equator, and the Meridional Ocean. And note, Genesis tells of "four rivers" from a single source dividing the land.

Perhaps, to the original inhabitants of Pangaea, their Pangaea is a flat island surrounded by Ocean. The other edge of the Ocean disappears behind the horizon and behind the worldly scene. The waters of Ocean appear to rise to create the dome of the sky. Hence, until today to many peoples the world is flat. But then convolutedly, does this not remind us somewhat of a hollow Earth?

The four continents of Krates are not the continents as we know them but are represented by Europe, as an isolated continent; Asia and north Africa as one continent; North America, Central America and northern South America as one continent; southern Africa with a few other southern lands as a continent; and southern South America with a few other southern lands are as a continent. The Aequatorial Ocean —note— would place the Congo of Africa, the Amazon

Valley of South America, and some of the Indonesian Islands under water. That at one time these lands were submerged is attested by their elevation today being a little above sea level, being some of the lowest lands in today's geography of the world, and perhaps the most threatened as the ocean levels today are rising.

Obstacles to the theory are the north-south Andes Mountains through Peru and the north-south mountains flanking the Great Rift Valley of eastern Africa, unless one considers that in both continents, these may have risen at a later date due to squeezing together and the clashing between continental plates, or that they have rotated in their positions, or that the Earth was more tangerine-shaped then.

– –––– –

Revisiting Pangaea, it would make sense that the Pangaean continent, impending to split into parts and to begin the great flow of the parts toward the equator, the splitting would not be random, as at that time the spinning and orbiting Earth was in balance and in a more perfect circular orbit. In Pangaea's uniformly stretching toward the north into an increasing circumference, Pangaea is incrementally stretched laterally until ripping into four parts. Why four? The ripping is short of uniformity and is quite vibratory, bespeaking a beginning tilt and the beginning of an "anomalistic quartering" of the Earth's orbit.

That the Great Continent Pangaea cracks into four pieces introduces a geographic and mathematic problem. The bending moment of the floating continents and the differentiating circumferences northward from parallel to parallel toward the equator inevitably leads to a tearing apart.

Imagine families and clans splitting as the land under their feet cracks, spreads, and separates them for an eternity, or at least until boats and then airplanes come to be. And they meet, but only to reject each other as strange cases for existence! In turn, all bespeaks the rise of four seasons, each becoming differentiated accordingly, each of a different temperature with different rates and limits of expansion and contraction. Diets on different flora and fauna yield a variety of human minds and physiques.

With the developing of terrestrial wobbles, the irregularity of continental shapes, and different types of human beings, flora and fauna come to be further complicated. Perhaps such violence induces the rise of violence for the incidence of unfulfilled needs, until Man is forced to begin communicating with "strangers".

Perhaps these are telltales of the Earth at one time having a shape more lemonoid, rather than tangerine-like, due to a changing speed of rotation about its axis, while the momentum of the waters for a spell had kept the waters closer to a spherical shape, which naturally would come to form the Aequatorial Oceanic Belt. The Meridional Oceanic Belt supposedly goes through both the Arctic and Antarctic regions, but it is reasonably widened to form the Atlantic and Pacific Oceans. The case of its going through the Antarctic may be explained as the void left in Pangaea's wake as being new matter coming up through the original springs, which eventually widen into the South Pole oculus and deposit new materials to form the present Antarctica. The Arctic Ocean today is somewhat constricted by the Bering Straights and a few Canadian Islands, although technically it still is a part of the Meridional Oceanic Belt. Or, perhaps, the Earth hosting Pangaea may have been flatter than a tangerine, which would have caused a more violent breakup of the continent.

And quite important perhaps, it suggests that Pangaea may have had slipped northward to the one side of the world, leaving a more simple explanation to the forming of the Pacific Ocean.

The Mercator's Chart of the World, Flemish, 1569, shows a considerably enlarged Antarctica with Australia attached, along with other islands.

The Piri Reis map, a Turkish copy of an older Arab sources, in turn copies from the ancient Great Library of Alexandria, Egypt, was shown to Christopher Columbus by a Turkish admiral. It suggests that the southern tip of South America was contiguous with Antarctica, that is, without the Straits of Magellan. And it shows the Antarctic mountains under the present-day ice cap quite accurately, which recently were discovered through radar. If so, can you imagine, with no Straits of Magellan to let water through, all the isolative ecologic repercussions? As ocean water levels unequally rise, current and fish migrations change. Adjacent lands dry up, sink, flood, or change in kind of productivity. Water evaporation and rain patterns are altered. People readjust to a new environment and go on living and progressing.

The scholar and cartographer Pompei, Italian, 1731–1788, speculates with the idea —or based on genuine research into certain lost sources— that the Indian Ocean was quasi-landlocked, showing that Indonesia was contiguous with Malay and Australia. If true, the "land locking" would have been only temporary; nevertheless, other ancient maps do not close the waterways between Australia and Antarctica.

There are ancient maps showing that the Caspian and Aral Seas were one body, and that together at one time, this body of waters was open to and a part of the Arctic Ocean. Perhaps there was a time that Europe and Asia truly were two separate continents, as maintained by ancient tradition.

Many ancient maps show an abundance of localized anomalies. Are these mistakes or recordings of the past? All these ought to be taken more seriously, as suggestive for research projects, rather than be belittled as short of accuracy or abjectly dismissed as primitive.

On Man's thoughts and activities, it is said, everything is written in the stars. And now we learn that much is written in our DNA (deoxyribonucleic acid, found in our blood) about our heredity and ancestors. Yet perhaps to be discovered are all the conversations each of our ancestors had and overheard may be recorded in our DNA. Students of the great pioneering electric scientist Nikola Tesla, Serbian, speculated that all our conversations, past and present, are recorded in the vast space of the universe – that is, the wavelengths are not dissipating. What an indifferent walking warehouse of records each of us is! And, sadly, we are not ashamed of our nonsense and our not knowing what our DNA knows about us. Tree rings give the ages of trees, but does not each ring, while in it's making, record all the sounds of nature and human conversations in its environment, besides evincing physically the varying weather conditions of each year?

Now, is it possible to hide from the future and keep everything of the past a secret? Ultimately, all becomes an open book for the Repented and the Atheists. The record of evil, however, for its arrhythmia and exclusion from divinity, eventually self-deletes from memory, in the face of rising universal harmony.

The body produces melanin, the coloring in the skin and hair of Man and in certain animals, at nighttime, while sleeping in the dark. Melanin protects against sunburn and related diseases. The darker the location, with the least reflected light, the more melanin is produced to increase the darkness of the hue of color in the skin. At the equator, where the nights are darkest, the production of melanin would be at its thickest. The color of melanin (in spite of its name), in small amounts from the least to the most, yields a range from light cream color to dark brown.

When there is no angle of precession to the axis of rotation of the Earth, that is, when vertical, the two polar regions are permanently in sunlight. The Sun keeps hovering around the horizon on an even keel. Its rays touch the polar areas tangentially and so are weakened, but longer time-wise. The snows reflect a bluish light day and night. Exposure to blue light gives the skin a "suntan" color and vitamin D (additionally, which is other than melanin).

When a small angle of precession occurs, which describes a limited Arctic and Antarctic Circles upon the celestial sphere, the six months of night within them are not dark, for the greater amounts of light reflected from the atmosphere. As the angle of precession increases, the Arctic and Antarctic areas increase, and accordingly they darken.

"Summer" or "winter" seasons, if any, would be a phenomenon, other than as known at present; "summers" could preemptively advance to pass an equinox and come to fall upon perihelion, and "winters" at aphelion (per the pre-classic Orphic poems). Aphelion, then, would yield yet darker nighttimes. Throughout his civilization, Man, by nature and by divine guidance, learns to accept and compensate for stranger climates to overcome his physical inadequacies. Through such, primordial migrations may be anthropologically traced.

The geographic parallel of habitation at some primordial time, along with other possible factors, it is theorized, had determined the body shape, kind of dexterity, and colorings by the biologic mechanics of the germane times. By providence of the Creator of nature, the purpose of such is for protecting against solar radiation and related conditions, as well as for differentiating the uniqueness of the ingenuity of each of the peoples and eventually to become complementary to each other, rather than competitive. With the shifting and fracturing into continents, theoretically the diversity in the human, animal, floral, etcetera, kinds in their isolation are more acutely differentiated by nature, thus in time encouraging trade, human interdependence and learning. In the field of sports, it is obvious how some ethnicities naturally and uniquely excel over others in certain contests, and others in other. Mankind becomes a garden of variety; Man is likened to all the beautifully colored flowers that Man tends and enjoys.

How about human life on and in other planets? Mythology says there is. Religion tells us of heavenly hosts. But the obstacle today is that we are trained to look down on Mythology, even though it is rather quite uniform among the many ancient peoples on different continents. God tells us that He created the entire cosmos (which specifically includes all that is suspended and moving in the celestial space) on behalf of Mankind. Now, who knows how to interpret this? If there are signs for a humble Mankind, Man is charged with protecting the environment, natural and animal, rather than voraciously indulging, consuming and wasting all, until attaining an imbalance, arrhythmia, and self-destruction. And he is charged with protecting outer space from pollution, too.

The languages mankind speaks are shaped by the foods he eats, the aesthetic and physical qualities of his environment, and the temperature range of his habitat. In the cold regions, so as not to lose calories, talking is through a slightly open mouth, and in the hot regions to ventilate, talking is through a wide-open mouth. Man may not improve himself with dietetics and exercise alone, but as the ancients knew, by following the rhythmic cycles of feasts and fasts, and the periods of rest and relaxation, as well.

— ——— —

Unlike animals, which react predictably to changes in nature, a human being actually and pleasingly may influence nature, provided he is clean at heart. Human beings collectively, whether they like it or not, or believe it or not, issue forth vibrations that can exacerbate negative conditions, or they usher in good weather with beneficial conditions, as natural consequences, besides their being Divinely answered. Such vibrations in religion are the human voices in the community prayers, litanies for rainfall, worthy public outcries, etcetera. Human beings may fool each other, as for sincerity to each other, but their vibrations fully will deny and negate their prayers when hypocritical: "Know thyself!" —Socrates. Example, a boxer looks his contender in the eyes telling him more quickly about his expected next move.

— ——— —

A Divine scenario of checks and balances appears in 1967. The fruit we eat are industrially "frankensteinized", de-foodized through pre-harvesting, gassing in warehouses for a season, and left only with tasteless fiber, and a deceptive coloring. Grains and vegetables are "frankensteinized", de-foodized and left with fiber and fat.

A movement for women's liberation breaks up families, increases the number of illegitimate children, abortions and the crime rate. At the same time the incorporated cattle, hog and chicken industries, and the farming corporations by "frankensteinizing" accelerate the growth of their products by feeding them hormones, toxins, soy, etcetera. Innocent consumers say, "I just simply lost my taste for eating meat." But, in Divine synergy, do not many girls, as a timely result of being fed un-naturally tampered meats and dairy products, mature earlier into womanhood with cumbersomely heavy breasts as a timely reminder to them and to all society of their Divinely pre-ordained motherhood and obvious womanhood?

— ——— —

60 — A PRACTICAL CONCLUSION

Tà pânta rheî. "Everything flows."

—Heraclitus

Or more incisively and paradoxically, a plural neuter adjectival subject with definite plural article and a singular verb, for the semantic nuances in the Greek language, in loose semantic translation, that would be: "All non-personal things and inanimate matter that may be, is in a state of flux." All (plural) that physically is, all that is created by the Creator, always is (singular) being in an inanimate or irrational state of flux. Or, grammatically ackward in English, "All things, flows." This apophatically would allow Man, which is Masculine, and his good to have permanency.

The cycles and changes in the relationships between Sun, Moon, and Earth do influence and are influenced by each other, and indeed with the star Seirios, and the rest of the solar system, and beyond our galaxy, all through a greater perpetuating government. Those same cyclings and changes do not go without influencing and changing the physical and biological forces down to within each cell, molecule, atom, and so on, of all Earthly substances and DNAs. Other planets and celestial phenomena are not exempt to reactions from what happens on Earth.

In eras of the long past, when the gravitational pull suddenly became greater, the alleged giants and dinosaurs perhaps collapsed and died off because of a mortal increase in gravitational pull on the massiveness of their bodies in proportion to the area of their footprints, or the change of oxygen supply, affecting their basal metabolism, or we do not know.

There are no absolutes under creation, not in motions; not in fixations; not in measures; not in weights; not in mass; not in directions; not in temperatures; not in speeds; not in rhythms; not in time, seasons, or periods; not in forms, neither in the measure thereof, nor in Man's units of measures; nor in the knowledge in full of any created thing. Even that which some call the allegation or theory of "evolution" needs to be redefined or rethought to remove its sophisticated adversarial barbs against the whole truth, the Divine Truth. It is said that the purpose for the "theory of evolution", as academically presented to humanity, is to prove that materialism is the basis for existence: it so fails.

It is known today from laboratory tests at McGill University in Montreal, Canada, that radioactive elements decay into inferior elements, an atom of gold into an atom of lead, a process called alchemy. The Roman Emperor Diocletian had led a force into Egypt to destroy scrolls in the Alexandrian Library on making gold. Gold is the most sanitizing metal known; when the time comes for all commoners to eat and drink out of gold table settings, food poisoning may have to find a new abode, per an earlier Christian prophesy.

How much does Man really know about the Earth he lives on? How much about the Heavens, from which the "Earth is suspended", per Solomon, in the Old Testament? Of all the heavenly bodies and all other galactic systems? Man's research perhaps is mere speculation, expressing only an imagined opinion. With each increment of revelation and successful discovery, Man finds manifold more of what keeps eluding him and of what he does not yet know. Thus he grows ethically, aesthetically, and spiritually along with the physical truths he discovers.

The Creator, who is the Creator of all that is, and is the Beginner of time, is the only eternal Absolute. All else pertaining to reality is relative, elusive, and in a state of flowing re-adoption, including our own corporeal substance and security. (Numbers are absolute. Other abstract concepts are apart from reality.)

Man is endowed with the characteristic of flexibility and adaptability. So is his Earth; or else it would long ago have shattered, perhaps into cosmic dust, in the face of gross changes, be they shockingly traumatic or ominously progressive or indeed good. Man can also constrict himself mentally when he chooses the "dumbing down" path. He may constrict his knowledge into the imaginative greatness of his isolation, or he may expand his knowledge in the reality of his greater humility to see clearly what little he knows in the vastness of His —yes, it is **His,** created for him— universe.

The Heavens and the Earth will react to him just as accordingly, either to confuse him, or rather challenge him, so he will investigate further for his benefit. When in harmony, the Heavens and Earth react with him.

The planet Earth cyclically changes internally and superficially. With each of the cyclings, its several movements are reshaped in configuration, activity, temperature, capacitance, and character. These are due to readjustments in relation to motion, space, time, and the electric potentiality lying in the vastness of the universe. All these are epochically manifested in the calendric changes. So Man's sense of time in his worldly environment does change, as does his developmental history. He no longer fears falling off his Earth; rather, he now may fear falling back upon it. And not without significance, the human scale likewise undergoes alterations, and not without occasioning differentiating experiences in Man's civilization.

The story of Man's civilization covers four most significant Great Ages of ideologic and moral development: the Golden, the Silvern, the Copper, and the Iron partially intermixed with Clay, or is it intermixed with Plastic? Each of the Great Ages may contain thousands of years, per Hellenic Mythology, and the book of Daniel, and they agree and clearly depict Man's changing awareness and activities. But modern science up to the end of the twentieth century tries instead to speculate on the Stone, the Copper, the Bronze and the Iron Ages of speculative and materialistic development. Man today fantasizes that he is smarter than ever. The Mythologic concept of the Four Ages is materialistically fugitive, and Daniel's concept (Old Testament) is materialistically accumulative; the first is spiritually working down to the applicative, and the second is materialistically working up Man's ego to his greater physical might, oxymoronically.

Rather than choosing sides, the wise researcher looks for the ever-crossing vectors to keep his current situationalism throttled and in balance, dividing between what is known, what seems to be known, what is not known, and what possibly could yet become known, rather than the die-hard attitude of having to make a decision, even when not prepared, even when not necessary, until comes the time. He must divide what seems practical for a short expedient term and what stands immutably practical in the long run.

The Earth is Man's present home. When he treats his home with love, his home offers him comfort. If he treats his home like an expendable commodity, his home Earth may trash him.

– ———— –

A phenomenon of mammals and human beings: When naturally sitting on the ground, on their haunches or cross-legged, the opening of the rectum sits squarely centered and directly on the piece of earth they occupy. It is as though plugging in to draw terrestrial or tellurian energies up through the digestive system and into the rest of the corporeal system. But when wearing clothes separating the contact, the question is, what is being denied. Or are there other perplexities, or what is enhanced when energy passes through a certain layer, or what particularly is being screened out?

By wearing shoes, or with shod hooves, etcetera, what is denied to these living bodies? Based on the science of human "foot reflexology", each part of each foot represents, stimulates, or receives through the nervous system a certain massaging therapeutic energy, or tells that a certain gland or organ is calling for help. Wearing all natural raw leather shoes, however, may be tantamount to going barefoot.

The old fashioned physician with a small flashlight looks into each eye of his patient, which indicates or mirrors the health or ailment of each internal organ. The iris of each eye records the history of health of each organ of the body, also the mental state, This science is known as "iridology". For curing or treating, possibly one day, the appropriate part of the eyes, or fingers too, will be bombarded with rays or vibrations to stimulate or repair a particular part of the body, the counterpart to foot reflexology.

Sunshine and moon-pull we take for granted. Do we know all the positive and negative influences of the many other stars and planets when looking at them without wearing glasses? Some celestial bodies disappear from view a part of the year; what are we being relieved of temporarily?

The stomach has 10,000 different kinds of digestive glands, which either alone or in combinations, receive all that makes its way into digestion. Can the rays or energies of each of the planets and stars make its way into a healthy digestion, say, when outdoors on a picnic? Can our bodies be remolded and tempered with such celestial influences? Are we caught into a celestial web and in need to free ourselves from unnatural entanglements, and to learn to walk through natural botanic environments, and talk and listen in healthy oxygen? Unfortunately, modern medicine is being forced to isolate Man's need for therapy from the strong power of his universe.

— ———— —

Many cultures agree, Satan enters human beings through their mouth, particularly at such times when the wrong words come out. Two eyes, two ears and two nostrils take in all that is good and bad, and one mouth to talk straight, and to allow forth only one sixth of the internalized out-going matters. And it is no Divine accident that thirty-six teeth surround his tongue to protect him from his own imperfection, to insure his discretion.

Intake of other food values and energies enter the human body as well through the skin and hair. Hence, there is the possibility for a superman and a super-brain.

If in all the universe there are the many relationships and affects with equal and opposite reactions, then does this not influence all of Mankind wittingly, unwittingly, or autonomically, not only to move mountains, but as well to reshape continents and oceans, to reconfigure the globe and its texture, and reorient with respect to celestial positions, and perhaps as well to affect all the remote planets and stars? Then, will he not be able to remold all, subjecting all to his sense in parallel with his thinking?

The recent oil well explosion in the Gulf of Mexico and the vast spill of oil floating eastward to the open Atlantic Ocean and turning northeastward has brought on an unusually severe cold condition in Europe. The sheet of oil insulates the warm current underneath the spill from releasing its warmth into the air currents. Likely, this is not by Man's will, but by Man's stupidity (or evil inclinations).

Nevertheless, the fuel industries in USA are losing business and the people are saving, and the fuel industries of Russia are gaining substantially in their European markets, while Russia's northeastern shores warm up as the oil sheet fizzles out.

– ———— –

As it is known that the sound of running water from the faucets cause gentlemen, not quite in health, to run to the toilet. Then not only music, but all the sounds in nature and the universe, as winds, waves, whispering leaves, bird calls, have their healing powers on us, bodily and psychologically. Fast trains, industrial noises and explosions may, instead, have adverse effects on us. Do the stars and planets have healing sounds for us, as subtle as they may be?

Petting pets is most therapeutic for human beings of all ages. Children used to love to pet the horses of the milk wagons and the horses loved it as they were doing their chore. Our industrial and automating revolution is estranging us even from the cheerful birds, which are becoming a rarity,

That is, as Man approaches his foreordained and ultimate perfection, the "Likeness" of God? For, when Man fell, all the universe fell, as is written. He needs only to stand up again and clean house.

METAPHYSIC
WORLD

Visualizing the worldly spaces and invisible structures between us and the Heavens, and colonizing these etheric spaces.

61 — CELESTIAL ICOSAHEDRON

The tree of pauses, which is not easy to refute, suggests that each of the atmospheric and other higher etheric layers may have some solid geometric, or a geodesic Portolano-like skeletal, or "celestometric", substructure, rather than simply being of random concentric spheres.

If by any natural solid geometric phenomenon of energies, as glare would indicate, such could answer only to a web-like framework in the solid geometric or geodesic form. Most feasibly it would be of an icosahedron (icosahedric, adjective). And if so, there may be a series of superimposing layers of such icosahedrons, each to relate to each of the etheric spheres, as though in a parallel celestial universe.

In solid geometric terms, any space may be three-dimensionally demarcated, either into Portolano-like triangular-based pyramids (each of four surfaces, including base), or into Cartesian cubes (each of six faces). The case at hand leads to the solid Portolano-like solution, which may be known as the icosahedron solution.

Icosahedrons are like twenty-sided dice. Or, such is as a diamond ball, upon which the diamond cutter chips twenty triangle-shaped faces, or simply a celestial crystal.

Technically, an icosahedron is defined as a full all around geodesic figure, which is composed of twenty equilateral triangular flat surfaces, twelve low corners and thirty open edges (strings), uniformly reaching all around. It is a rigid self-supporting structure, capable of rolling around.

Or, more analytically, an icosahedron is formed of twenty triangle-based pyramids, all of which are drawn together, each at its apex, fittingly; and the joined apexes form the icosahedron's centroid.

In any of the foregoing cases, each set of three triangular faces on the icosahedron's surface forms a low pyramidal corner. All of such corners serve as nodal junctures of each three surface edges and a fourth hidden edge (string) to the centroid. The centroid may be taken as a hidden thirteenth corner, another nodal juncture.

— ———— —

In orientation: A four-surface pyramid (triangular based pyramid) gives four directional lines equally askew to each other, at 30 degrees. A cube gives three directional lines, x, y, z, at 90 degrees to each other. And an icosahedron gives ten directional lines equally askew to each other, at 30 degrees, as each directional line goes through the center of the two opposite faces and through the centroid. All this calls for new mathematic systems.

In constructing an icosahedron, 30 strings are required. Or, 12 more, if to connect to the centroid, though structurally not needed, then would total 42 strings. All angles formed in a four-faced pyramid, and in an icosahedron, regardless of orientation, uniquely are of 30 degrees. Structural engineers come up with many varieties of rigid self-supporting geodesic domes, including with subdividing each triangular face into smaller low pyramidal triangles, but all these, although on the triangulating principle, depart from the isosceles triangle of the exclusively 30 degree characteristic of uniformity in all directions.

Each of the icosahedrons, engulfing the etheric spheres and the Earth, may rock rhythmically to and fro, and vacillate with each other, as well below the Earth's surface. These icosahedrons theoretically and technically may lie across the pauses or across the etheric spheroid layers, or tangentially cutting through them as necessary, or with bending sides to follow the bowing contours of the pauses. Imagine an icosahedron of 30 stiff elastic bars (strings), which when bounced, springs back straight to its original proportion?

— ———— —

Let the centroid of the icosahedron be the centroid of the Earth; and let the Earth be a sphere contained within the icosahedron. Thus, established are the criteria for three-dimensional navigational and engineering charts with measurements based on the "Portolano" triangulating principle, and to include verticality, a "solid Portolano" principle, by which to establish the structural nodes and strings of energies, as far up to and past the magnetosphere.

The axis of rotation of the Earth, as set, let us assume, goes through the center of the two opposing triangular faces of the icosahedron to avoid complications with the trunk of the tree of pauses; and the equator straddles, while cutting across through six triangular faces, alternating, three upright and three inverted triangles. (That is, none of the three-faced nodal points lie on the equator, or on the axis of rotation.) These nodal points serve as the basic points for the triangulation pattern for subdividing the Earth's surface with triangles.

(Theoretically, the icosahedron when in a tilted position, by 15 degrees, would have two of its low pyramidal corners lying on the axis of rotation and the other ten straddling the equatorial plane. Although worthy and valid, this latter possibility is not feasible for this adventure. However, at some primordial time, when the Earth's axis of rotation was upright, with two of the nodal points on the axis, the projected division on the astral belt would be of ten zodiac signs.)

The twelve nodal points, as projected upon the Earth's equatorial plane alternating from either side, straddling the equator, and sequentially projected upon the equatorial plane and radially outward, relate directly to the zodiac belt; but with a slight differentiating slippage each year (hence, the Juliano-sidereal great year, on which see Chapter 52).

The prehistoric peoples had known of the earthly triangular nodes, composed of electromagnetic nodes (knots) on the Earth's surface, and of their very serious effects, such as for their keeping insects away, their psychologic influence upon those found present, and as grounding points (transformers) for receiving and disseminating celestial energies at these locales, and perhaps having communications of sorts. Upon such electromagnetic nodes the ancients build their Pagan temples, oracular centers and shrines. The early Christian, and the Medieval churches in Europe and the Latin Americas, as so far verified, are built on these same ancient "sacrosanct" nodal sites or foundations on the Earth's surface.

Dried up marshes naturally give us a clue, they parch up into hexagonal patterns, as famously in northwest Ireland. A hexagon is of six triangles joined at their apexes. A hexagonal pattern most feasibly allows for expansion and contraction of material matter, allowing for the least distortion. A parched flat surface of a hexagonal pattern, when expanding, and being forced into a convex curvature, into a spherical segment, naturally will split into six pieces of a pie. By being stretched at its weakest point, its centroid, it opens into a low lying hexagonal pyramid with creases splitting toward the opened out top, all within the limits of the hexagon's six sides. Perhaps, the crystalline web of electromagnetic forces has the major effect for the hexagonalizing of a parched surface.

— ——— —

The sharp reflection of sunlight upon a smooth shiny surface in the open air produces a glare, asteroid in shape, with clusters of rays faintly appearing hexagonal, which bespeaks further of an icosahedric natural law. (The glare through window glass, or plastics, or lens-like objects may come out differently, as perhaps cross-shaped.)

Crystallization, depending on the environmental polarization, may be hexagonally symmetric in two dimensions, and cylindric in the third, as quartz. And when free of polarization for any reason, it is symmetric with respect to three dimensions, as are snowflakes. This latter rule may also apply celestially, by the "icosahedric" method, as when endeavoring to allow atmospheric and exospheric energy to attract a physical buildup of durable materials of tensile strength around a string of energies, to form a durable structure in celestial suspension.

— ——— —

Also, there takes place a field of mingling of energies of other kinds, not yet investigated. Energies of other kinds may be likened to the human anatomy, its vascular, nerve and lymphatic systems.

These most likely would offer directional energy activity along the Earth's axis (trunk of the tree of pauses), as a conduit or core of energies, which also may hold a memory bank of all that takes place on Earth since the beginning of time. Extraterrestrial intelligence, would there be such, may long ago have tapped into our private and state secrets of the last twelve millennia.

Besides the many forms of electric currents and energies, there may be unlimitedly more kinds, to include cohesive, adhesive, latent and hibernal (which defy detection). Some of these may aid in the metamorphosing of dust and particles into tangible physical lightweight (or anti-gravitational) matter having characteristics of tension, compression, and bending, and vibration to withstand terrestrial and celestial undulation, and heat and cold transference, and to include memory. (Nikola Tesla, 1656 to 1943, claims that all conversations remain registered in the atmosphere.)

Is, then, each galaxy in the universe, composed of a random collection of more closely related stars and planets? Or, is there some greater energy shell, which specifically contains each of the galaxies in a honeycomb-like energy distribution, or as a "hypercosmic" atom?

With the Earth's equatorial plane cutting through the greatest star of our galaxy, Seirios, then the tree of pauses must point perpendicularly to a location of comparable celestial significance.

Some wavelengths must be faster than the speed of light. Perhaps the aurora borealis and aurora australis are lights of phenomena of changing wavelengths transitionally going through a visual frequency, while straddling the axis of rotation and crowding through the polar oculi. Alternating currents and direct currents rhythmically shift from a general direction, from an infinitesimally short time to many years. These also may accede the speed of light, as born out by the extra-sensory perception by animals and by certain human beings. Upon passing through the galaxial shell these currents refract into the fabric of the universe, precipitating and depositing the memory of continuing natural and conscientious thought, as small as we are. And perhaps, these may serve as "blue prints" for random human thought and "deliberate" activity, and for animals to do what animals do. Do these affect other oscillations here and beyond? These phenomena may serve to explain why some individuals think that at some time in the past they were some historic personality (besides the genetic descendant theory).

By astronomic observations tracking down the speed of distant galaxies moving across a telescope, the speed of some is determined through triangulation to be faster than of light.

– —— –

The polar holes of some of the terrestrial, atmospheric, etcetera, shells may occasionally line up, but for all to line up approaches an improbability, as all terrestrial and celestial motions are accompanied with reactive undulations (as in third note tunes, and generationally more between each two notes in the series of generated chords). The pauses react to such adjustments fluidly without friction, and with ease channel off their particularly accrued capacitances.

When the poles are clearly and sufficiently open to the atmosphere, does a clear ethereal channel from outer space open up along the closer confines of the axis of rotation, focusing and leading into, through, and out the other end of our world? Stretching the imagination, is it like an express tunnel or shaft in a so-called parallel world cutting through the Galaxy to beyond? Is it like a tree with branches above and below (roots being the underground tree), where typically following midnight waters are being transported upward, and following midday oils are being transported upward, to nourish the tree, and alternating, to nourish the roots likewise from etheric matter? However, can the energies in a tree trunk penetrate through a thinly closed-for-the-duration oculus? Accordingly so, yes, perhaps muffled.

Would any similar channeling introduce and remove different kinds of matter and energies from the Earth? Would space travel eventually become feasible, with frictionless "blastoffs" (or, say, suctioned upward through an induced lifting vacuum or by anti-gravity auto-levitation) and cushioned landings?

It makes sense that the gravitational field and the electromagnetic field are not the same, per Timothy Leonakis (see Chapter 64 – Addendum), to whom therefor the undersigned is indebted. As he presents it, one more argument may be deduced for an Earth Hollow.

Assume two cones with their apexes pointing at each other at the center of the Earth, projecting through the polar openings, opposite each other (similarly to Figure 11-H and 11-I), on to the opposite limits of the Galaxy. The energies perhaps are in one celestial direction, being grounded, recharged and bounced back from one side of the Galaxy, through the Earth and on to the other side of the Galaxy, cyclically. Or, they are being bounced through another planet or star, like playing celestial "billiards", thus establishing a web of trails of memories, so to say, a solidification of thoughts and deeds, serving newly amalgamated energies to accommodate expansions of all sorts, including timely human enlightenment. Perhaps there are connecting fibers of a sort, between other galaxies. The possibility for many phenomena may be mirrored in history being devulged by the rings in a cross section of a tree trunk. Perhaps the tree rings of trees, technology permitting, may reveal the sounds and pains of nature, animals and human beings like a "victrola", including of petrified trees. All sorts of terrestrial and rocky stratifications may well contain the sounds of the time of their stratification.

Is it possible that the Galaxy contains a vast intangible electro-energy tree, or say, twelve or twenty trees, each with a root onto a certain star, as perhaps onto Seirios, from the branches of which depend the tangible planets and the rest of the stars, as decorative lights and ornaments on a Christmas tree? This would further satisfy the enigma of an icosahedron. But, whatever Man so far has not detected, may be classed as speculation, but may lead —even by some insightful cartoonist— to yielding some fruit. If it comes for matters to be so, then we may approach the new matters at hand with a more level head (and shackle up into a safety deposit jail our "hyper-practical-ninety-day-return kind of "politicians", as dangerous animals).

— ——— —

Let us speculate further on another plane of imagination. Suppose a collection of supra-terrestrial energies are distributed in a network of triangular nodal points, linear, radial, spheroidal, icosahedric, etcetera forms, say, at the limits of the atmosphere, or higher yet. Such nodal points, stretching our imagination, may neither be supported from below nor suspended from above, but be floating anchorage corners of a parallel universe.

Between such nodal anchorages, there would be found the inactive quiet areas and the most turbulent areas. All such would form a network encompassing the Earth. Or, as though an icosahedronic celestial crystal, which encapsulates the Earth, and with rays which refract pyramido-triangularly (instead of simply prismatic). Such icosahedronic crystal would facilitate the establishing of the solid Portolano-like points of nodes of capacitances on the Earth's etheric sphere, as on the Earth's surface, as early Man had indeed discovered to build his temples.

Little packets of energies, from docile to wild, may be found in the network, as even in our homes, in simple suspension (Lynn Taggert), or throbbing, or cycling, or erratic in various globules or unusual forms, as in the forms of cubes, kidney beans, pyramids, little spirals, and perhaps into more complicated forms, as undulating jellyfish, electric eels, or in images of flora and fauna, as found on the zodiac, and on the Earth, especially in the ancient art work in the deserts of Peru.

The standard measuring for fixing points and movements in the celestial sphere is by the simple 360-degree horizontal and 90-degree up and down system, or the simpler Cartesian "X, Y, Z" system.

But as well, the pyramidal triangulating Portolano icosahedric system may be given a try; it may initially be complicated, but of a use for whatever yet to be established. An icosahedron may be projected infinitely into the celestial sphere, using measures as celestometric proportions (in allusion to geometric proportions), rather than as quantitative digits.

— ———— —

The axis of rotation of each etheric layer and its matching icosahedron (in the case of multiple icosahedrons) as a celestial couplet may share a common axis of rotation. The equatorial orbit of each couplet may relate to other particular star, or to a couplet of stars in our galaxy.

— ———— —

62 — DIVINE PRE-EXISTENCE

There is a difference between a straight-line evolution and a periodization into "days", per Genesis. "Day" etymologically is a period of "rollover" of time and accomplishment of a certain activity. "Day", in Greek 'Hemera', as used in the Septuagint, and as in the etymology of the word, is the nature of a genuine evolution, but that would be within each of the subsequent days of Creation, a genuine "turning forth", or "turn over", or a "running around and over" is brought to perfection and completion with each cycle of a period called a "day".

That the Earth is growing from the watery bombardment and sprinkling from outer space would mean that the mass of the Earth and all the Heavenly bodies are in a state of growth and refreshment. Does this mean that this Day One, is an intended solution to the overpopulation and the shortage of food supply? Will Man stay at his present size, or shrink, or will become giants as of old? Not so fast, for there is some matter, however, being ejected through the poles, counterbalancing somewhat. Will this growth in diligence continue until Earth and the rest of the planets and galaxies are come to meet, gradually to touch and to fuse concretely to be integrally a single galaxy-planet? And if so, will there be a new understanding of infinity, such as: nevertheless, it is just as infinite and unknown in this new comprehension?

When Jesus Christ on the Cross, just before his Expiration, says, "It is finished" (John 19:30), is it that the work of the Sixth Day finally is come to an end? Is Man who is created in the Image of God beginning to grasp the advantage of coming closer toward attaining the Likeness of God, as promised in Genesis?

The very ongoing growth of the universe, as astronomic scientists tend to agree, bespeaks that there is perpetual motion, and that it is the most natural of the hidden forces in the nature that God has created for Man. Until Man comes closer to his ultimate maturity, without the deleterious help of the mysteriously materialistic cult of bankers, descendents of the builders of the Tower of Babel, and he will live free from distractions, then Man will substantially come to know.

As for the Seventh Day of God's rest, is it within the Day One? Yes, because it, too, is enumerated in the frame of Creation. (For, should physical time be terminated, Creation will cease to be; the rest of the matter would be theology.)

— ————— —

The **Lord's Prayer** is ephemerally translated according to the worldview of the Middle Ages; several versions already exist in English. A more correct translation of certain passages out of the Greek, affecting the matters at hand, is:

"Our Father, **Who art in the Heavens**, hallowed be Thy Name; thy Kingdome come; Thy will be done on earth, **as it is in Heaven**.

The, **Who art in the Heavens, "ho en tois ouranois"**, is plural, with definite article. The, as in Heaven, "os en ouranoj" is singular, without article.

The questions are: Are we being informed that as of yet the rest of the created Heavens are not peopled with rational beings like us, but that the Heavens are there? Why, then should we know of this? What does this mean in the light of our being allowed to bridge over between the celestial ornaments? Nevertheless, does His kingdom in Heaven, singular, mean, in general throughout the universe, but not yet on Earth with us? —As He did leave us to be, by our own activity and our will on this planet to cooperate mutually and to perfect our Humanity, as He has perfected the rest of His creation for us, as He is perfect? Note, not by leaders, but by the masses.

— ————— —

The Massoretically revised concept of evolution pertains to a straight line of changing conditions, but altering in accordance to

249

certain periods of Divine interventions. The Atheistic concept also is of a straight-line evolution, but smoothly through ever-changing conditions, ever-negating the present in favor of the coming of "new creations", which in reality is progressive anarchy.

If the Latin etymology of "evolution" means a "turning out", as it does, then the Massoretic and Atheist concepts of time and evolution are wanting. Or, they have chosen the wrong Latin derived word "evolution". The presentation of the entire work at hand is a development principally on cycles and spirals.

On arguing the matter of God's existence: Does material existence preexist spirit, or does spirit preexist God? Let the axiom be: God is Spirit and that He preexists, and that out of nothing He creates all that there is. Axiom closed. Really, the rest of Faith can only be based on hope. It is just as the theorems and proofs of geometry, all are based on certain beginning axioms and, nevertheless, they work for logical solutions.

Then as well, it is the Spirit of God, which potentially is everywhere and filleth all things, and energizes all energy, force, power and continuance of being.

The personality of God is not known to Man accept through the manifestation of His Son on Earth, Whom we see, hear, watch His work and what He tells us of His Father in Heaven.

– ––––– –

Oppositely, an Atheist may try to construct the universe as he would an aunt hill, higher and higher until his "Ant Hill of Babel" collapses from its own weight. An Atheist also may believe that natural laws bang into existence empirically; that may be excessively skeptical, just too imaginative.

The Atheist variety of scientists talk about Big Bang as being the abrupt beginner of all that is; and thereto somehow, inexplicably and in a change of tempo, they proceed to tag on their theory of evolution. Then, they present another stage, the random building block stage.

The Big Bangists by default are not too wrong, but they do not realize their own trap, too fast with the mouth, on the process of the "Bang" proper? Is the explosion and implosion cycle, a nanosecond, a year, millennia, as "Bbbb-bb-ba-a-a-a-ang-ng-ng!!! !! !", expanding all through the entire "Day One"? And all still is in the ever-active process of such a Bang at hand.

From the supposed "Big Bang", or whatever else one may call the ever-lasting "Day One", we can easily surmise that the second through the fifth days are eventful endeavors, each culminating into historic fruition of the so-declared intended accomplishments —"Let us....", of that particular day, and they are said in the past tense, which means that what is additive thereto, is not a replacement, not an improvement, but a new chapter, while cumulatively ans sequentially are ever-edifying. The sixth day remains so far today incomplete. For Man, being in the Image of God, has not yet attained the Likeness of God, as is projected in the Divine Intent. How misled and wrong are the New Agers!

Scientists who gloat in attacking the Creation per Genesis, contradicting themselves, announce authoritatively that the universe and our Galaxy are in the process of expansion, and will continue so. By default, again, those non-theologized good little boys unwittingly are defending the definition of Day One of Genesis.

— —— —

Unfortunately, the more "civilized" Man is becoming, the more monstrous he is turning out to be. Seeing himself, though short of merit, but as already "by his rights", as though he already be a divine likeness, as already being a god per the New Age hubris. That is, having eaten the apple of knowledge before attaining real and eternal life; he already knows all, that is, per the dumb-down paradigm, to all he needs to know. Such already holds himself irresponsibly above morality and aloof of justice, not yet knowing the virtue of humility, and as a collective, veritably a monster.

If there are plenty of crazy people on Earth, why cannot crazy planets and crazy stars be out there, too? Discords always dissipate out of existence, while harmonics multiply.

— —— —

63 — WHY DIVINE TRINITY

The following is not intended to be "theological", but rather, a reflection of His image, from what can be gleaned form His Creation.

For bringing up matters pertaining to revealed Trinitarian Religion and Mythology herein, the author apologizes: To the best of his knowledge, he uses certain unconventional arguments to present a fuller picture of a world view, as from the infinitesimal to the infinite and, conversely, from the infinite, through our human scale, to the infinitesimal. Man is in the image of all Creation, which he mist edify, and of God.

— ———— —

No one has seen God. Some have witnessed a manifestation of God, but not seen Him to know Him, but only to fear Him, and unquestioningly to hearken to His Law. Some speculate on the possibility of His Being through morality and rationale, and are convinced, that is the Way. Others, allegedly, have met one claiming to be Him in a dream. Others do not know Him, because of being limited to the range of their five senses. Others do not care. The personality of God is not known to Man accept through the manifestation of His Son on Earth mingling with Man.

Can any one word describe Him? If He is the All Knowing, what is the purpose and is anything done, and for what purpose? If He is Mind only, do we put Him in a glass jar in a preservative to star at Him for His Grace? If He is Purpose, an objective, a goal, how is it for anyone to point to Him, who is the subject setting Him off on such objective. or from where did such come to be pointing at a purpose, or goal, and how so mechanized? If He is spirit alone, what for and how, a spirit without Logos, or plan and direction? But, as the Mind, the Logos-Purpose and the Spirit, there we comprehend a perfect totality of the Divinity. That is, a: Be, Know, Do!

Man fell at some primordial date, and in his awareness and shame, he began to rise to save his family and tribe, then he began to see his security by examining his neighboring tribal tenets, and upon screening much by reasoning, he comes to an impersonal philosophic state. To be saved from egotism, and imperfect social goals, and indeed from usurpers God appears to Man, to the witnessing collective of Mankind (not just as to individuals), His only begotten Son the Word Jesus Christ Who shows us Truth and Love. Only begotten because He will not have another Son to replace what He has built, demonstrated and told us.

— —— —

In the book *The Disposessed Majority*, by Wilmot Robertson, 1981, the author makes a most interesting comment, that even lies, fraudulent statements and perjury in history, literature, government reports, all are vitally useful, as when one is able to look into and interpret the intentions for the deliberate misleadings, and a clue to what is it that is being hidden, or striven for clandestine advantage, he may read and see right through all such. Nothing, really, can ever remain hidden. "Mother, Why isn't the king wearing clothes?" says the little boy at the parade, as the king confidently is strutting off with the band. —Hans Christian Andersen.

Unfortunately, the more "civilized" Man is becoming, the more monstrous he is turning out to be. Seeing himself, short of merit, but already "by his rights" as though he already is a divine likeness, already is a god per the New Age hubris. That is, having eaten the apple of knowledge before attaining real and eternal life; he already knows all, that is, to the dumb-down paradigm, to only all that he needs to know. Such already holds himself irresponsibly above morality and aloof of justice, not yet knowing the virtue of humility, and as to his social collective, veritably a monster.

— —— —

64 — AFTERWORD, ON GRAVITY

Timothy Leonakis, P.E.
Bolingbrook, Illinois
February 20, 1995

Dear Nick:

With respect to the two topics that you have recently printed, if you would suffer a few comments and questions which these ideas raise, as stated below.

On the topic of the hollowness of the earth, there is an area, which is only slightly touched upon. Let us examine the influence of one of the three identified fundamental forces of the universe. Those three forces are nuclear forces, electro-magnetic forces, and gravitational forces. The force under examination is the gravitational force.

The observation that is made is on the gravitational forces and the application of its projected behavior under increasing field strengths. The popular belief of its behavior in concentration stems from the behavioral pattern in human cognition. That is to say when an observation is made and two witnesses, or data points are gathered attesting to the trend of the observation, a deduction is used to grasp hold of an understanding how the continued trend should be.

Deduction is a human trait that is used as a substitute in the absence of consciousness. How the torch of consciousness is ignited within a person stems from his awareness. Awareness is forced upon those who faithfully deduce since, as has been stated in your topics, there is ever change. The human trait of deduction allows the mind to go forward to the realization that deductions need revising during the attempt to grasp and understand the creation, which as also stated in your topic, but such deduction cannot be complete.

Now the deduction in place, which presently plagues the consciousness of conventional men of science, is that the gravity field, being observed to stem from mass and increasing with collection of mass is this: the gravitational field will continue to increase with increasing concentrations of mass.

As is with all deductions, this necessarily will break down and need revising, or will need to be scrapped altogether.

Now there are alternative theories for concentrations of the three fields, some of which are already known and demonstrated.

Of the three known fundamental forces, nuclear, electro-magnetic, and gravitational, let the case of the electro-magnetic field be brought under consideration. Man has been motivated for concentrating magnetic fields, so that transformers and inductors may be reduced in size, weight and efficiency. All matter is able to concentrate a magnetic field greater than that of free space. (But none has been found yet to de-concentrate it, or to carry concentrations less than that of free space.) The magnetic field is concentrated by coercive electric force to several times and even several thousands times its level in free space, as is in the cases of manganese and zinc ferrites. Then with increasing coercive force, the magnetic field suddenly drops and the material under intense coercive electric force behaves magnetically as free space, or as though it were not there at all in response to the coercive level. When the coercive field is reduced, then its magnetic behavior is recovered to its original concentrating properties, dynamically as the reduction in coercive field occurs.

This well-known field property allows credibility for the following idea: The gravity field of the earth increases as the centroid of mass is approached, traveling from its outer boundary toward the centroid. Then there is a sudden drop in gravity field concentration to that of free space for mass contained beyond the saturated gravity field level. That is to say, the mass cannot contribute to or sustain any additional gravity field beyond the saturated level and appears to be not there at all. In the saturated region, two separated masses are not attracted to each other, or anything else.

To date the only serious study of gravity fields has been in the case where the fields are weaker than that of the surface of the earth, or in a hole in the ground. No data is available on the behavior of gravity under highly concentrated conditions.

In the case of the theory of super concentrated gravity fields in black holes, which is unverified, there is no knowledge to say that body collapses from its center. That is to say a vacuum is allowed for the argument, that it is the surface which collapses, which it may not even do at that; possibly the mass has become gravitational entirely within a saturated domain.

Nick, this should summarize what is known today on the subject of energies.

Sincerely,
Timothy Leonakis

— —— —

OTHER WORKS

ONENESS OF POLITICS AND RELIGION

A de facto relationship
is dissected in the light of Patriarchy and Matriarchy,
and when which governs which.

A historical retrospective on the evolution of religio-political Man from time of Adam until today. The two most important aspects of Man's development and his continuance are politics and religion. In a historical sense they have most always gone hand in hand, and actually still do today, regardless of the modern liberal striving for separation of church and state.

— —— —

CONSTANTINE VERSUS THE BANKERS, *

Military-Industrial-Church Complex,
New World Order,
Today's Socio-Politico-Economo Fizzle,
& Big Dumb Down Conspiracy.

An epic of what went wrong and still goes wrong in a seemingly Christian setting, it examines what is yet to be accomplished toward Peace in this world. There is, nevertheless, a de facto oneness in politics, religion, education and public ethos, albeit, as four distinctly separate Estates, in the Byzantine Empire, and not always in harmony. This results at times in human failure and more often in perniciously working conspiracies.

Emperor Constantine still is maligned, from contemporary world leaders to historians, while Christians still are selectively killed with evermore imaginative cunning. Consider the sources vilifying Christianity: they are the same agencies hunting down the Apostles on their Mission for Christ, with the same combination of Murky Forces and modern internationalist adepts.

—   ——— —

JUGOSLAVIA AND WORLD PEACE *

—Vatican's Wars To Restore & Expand Caesar's World Empire—
—Alliance With the Christ-Killing International Bankers—
—NATO & Military-Industrial-Church Complex Eying Very Rich
Deposits—
Serbian military steadfastness humbles them in the Kosovo War.

The International Court, early 2007, at Hague finds Serbia not guilty of genocide. This means, the Serbian hero President Slobodan Milošević, who is officially murdered by poisoning in prison at The Hague, too, is not guilty of any acts of genocide and any related crimes as "berserkly" propagandized through official government and religious statements. This further humiliation of NATO was kept out of the News Media! His sudden jail execution saves several world leaders from embarrassingly facing international war crimes trials and life imprisonment.

The erroneously touted "age-old animosities in the Balkans" is a fiction of age-old enemies. The conflict in the Jugoslav Lands and the rest of the Balkans today is a mini pre-image of the coming Great Secular War III.

—   ——— —

OUR DIVINE LEGACY

—Christ & the International Vipers—
—Helleno-Christianity Versus the Figment of Judeo-Christianity—

Described is the methodical infiltration of the agents of the International Banking Nobility (Generation of Vipers) into the priesthoods of the pagan temples from Babylonia through Rome, then into the Temple of Solomon, and ultimately into forming concordats between the International Banking Nobility and the Club of "Christian" religious leaders.

By one such concordat a symbiotic relationship, Vatican's Pontifex Maximus owns all the souls and land of humanity and the Judaic Bankers own all the human bodies and gold. By similar concordats Freemasonry, Vatican, Zionists, Protestants and Muslims come under the dark wings of Big Brother. The scene of "the whore riding upon the beast" is coming into clearer focus today with the exposure of the "Military-Industrial-Church Complex". Greed for prestige, power, money, hate with big dividends, bountiful control over others, etcetera, surprisingly envelopes and taints politicians, clergy and bankers.

The scheme might succeed if philosophy and Classic Hellenic studies could be totally destroyed, and modern education "dumbed down" some more. Who, among several possibilities, will be "the whore riding upon a beast"? Where is the legacy of true Helleno-Christianity in the Name of Jesus Christ today?

[politically incorrect]

— ———— —

JULIAN CALENDAR VALID *

—Gregorian Calendar a Scientific Hoax!—

There are four kinds of calendars. Three of these have a specific ephemeral use. The anomalistic for wind, currents, natural migrations and cold and warmth. The sidereal for astronomy. The equinoctial (Gregorian) for fiscal matters. But none of these fit any of the inter-cyclings of times and conditions, and Man's body rhythms.

The Julian calendar (of very ancient base) encompasses Man's agricultural life and fits with the vital natural terrestrial and celestial cycling rhythms.

(Presently being revised.)

–– ––––– ––

THINE HEALTH

Your Own Personal Health! Cyclical Feasts, Fasts, Xerophagia, Food Abstentions, Therapeutic Baths, Exercises, etcetera.
By N.G. Phystiklakes & N.C.E.

–– ––––– ––

GYPSY COUNSEL

—Wheels of Serendipity—
A nomadic society urbanizing faces violence, trauma, romance & peace.
There is no Gypsy of fame.

The traditional Gypsy race or culture of people live in primordial isolation to preserve their ancient values, limiting their encounters to services and necessities among the strangers in whatever nation they live. They hail from no country, which they may call their own. This is a novel on life in America of social immunity and judiciary inter-dependence through the Roaring Twenties, Flower Children, Boomers, and Forever.

The Gypsies begin encountering changing obstacles by the early twentieth century, circumstantially being forced into urbanization, and find it difficult to protect their isolative cultural immunity.

Following a period of violence, they come to adjusting by accepting what is deemed good from society at large and abjectly rejecting all evil, as new means of cultural isolation and mobility are formulated to continue their unique survival.

The salient characteristic of their society is the total rejection of the "ego". There are no "me's", no "myselves" and no socially idolized personalities of importance among them. Their culture heroes, once a task is done, are thanked; and they fade away among the nameless "wise men" of the past. Only the beneficent works inuring to their collective are remembered.

— ——— —

www.ingramcontent.com/pod-product-compliance
Lightning Source LLC
Chambersburg PA
CBHW031828170526
45157CB00001B/221